SPILLS OF EMULSIFIED FUELS

Risks and Response

Ocean Studies Board
Division of Earth and Life Studies

Marine Board
Transportation Research Board

National Research Council

NATIONAL ACADEMY PRESS
Washington, DC

NATIONAL ACADEMY PRESS • 2101 Constitution Avenue, N.W. • Washington, DC 20418

NOTICE: The project that is the subject of this report was approved by the Governing Board of the National Research Council, whose members are drawn from the councils of the National Academy of Sciences, the National Academy of Engineering, and the Institute of Medicine. The members of the committee responsible for the report were chosen for their special competencies and with regard for appropriate balance.

This report and the committee were supported by a grant from the United States Environmental Protection Agency and the United States Department of Transportation-Coast Guard. The views expressed herein are those of the authors and do not necessarily reflect the views of the sponsors.

International Standard Book Number 0-309-08301-X

Additional copies of this report are available from:
Ocean Studies Board, HA470
The National Academies
2101 Constitution Avenue, NW
Washington, DC 20418
202-334-2714
http://www.nas.edu/osb

Copyright 2002 by the National Academy of Sciences. All rights reserved.

Printed in the United States of America.

THE NATIONAL ACADEMIES

National Academy of Sciences
National Academy of Engineering
Institute of Medicine
National Research Council

The **National Academy of Sciences** is a private, nonprofit, self-perpetuating society of distinguished scholars engaged in scientific and engineering research, dedicated to the furtherance of science and technology and to their use for the general welfare. Upon the authority of the charter granted to it by the Congress in 1863, the Academy has a mandate that requires it to advise the federal government on scientific and technical matters. Dr. Bruce M. Alberts is president of the National Academy of Sciences.

The **National Academy of Engineering** was established in 1964, under the charter of the National Academy of Sciences, as a parallel organization of outstanding engineers. It is autonomous in its administration and in the selection of its members, sharing with the National Academy of Sciences the responsibility for advising the federal government. The National Academy of Engineering also sponsors engineering programs aimed at meeting national needs, encourages education and research, and recognizes the superior achievements of engineers. Dr. Wm. A. Wulf is president of the National Academy of Engineering.

The **Institute of Medicine** was established in 1970 by the National Academy of Sciences to secure the services of eminent members of appropriate professions in the examination of policy matters pertaining to the health of the public. The Institute acts under the responsibility given to the National Academy of Sciences by its congressional charter to be an adviser to the federal government and, upon its own initiative, to identify issues of medical care, research, and education. Dr. Kenneth I. Shine is president of the Institute of Medicine.

The **National Research Council** was organized by the National Academy of Sciences in 1916 to associate the broad community of science and technology with the Academy's purposes of furthering knowledge and advising the federal government. Functioning in accordance with general policies determined by the Academy, the Council has become the principal operating agency of both the National Academy of Sciences and the National Academy of Engineering in providing services to the government, the public, and the scientific and engineering communities. The Council is administered jointly by both Academies and the Institute of Medicine. Dr. Bruce M. Alberts and Dr. Wm. A. Wulf are chairman and vice chairman, respectively, of the National Research Council.

COMMITTEE ON SPILLS OF EMULSIFIED FUELS: RISKS AND RESPONSE

JACQUELINE MICHEL (*Chair*), Research Planning, Inc., Columbia, South Carolina
JACK W. ANDERSON, Columbia Analytical Services, Vista, California
CHARLES F. BRYAN, U.S. Geological Survey and Louisiana State University, Baton Rouge, Louisiana
WILLIAM LEHR, National Oceanic and Atmospheric Administration, Seattle, Washington
MALCOLM MACKINNON III, MSCL, Alexandria, Virginia
JAMES R. PAYNE, Payne Environmental Consultants, Inc., Encinitas, California
GARY A. REITER, Westcliffe Environmental Consultants, Inc., Westcliffe, Colorado
JOHN N. SACCO, New Jersey Department of Environmental Protection, Trenton, New Jersey

Staff

DAN WALKER, Ocean Studies Board, Study Director
JOANNE C. BINTZ, Ocean Studies Board, Program Officer
KRIS HOELLEN, Transportation Research Board, Senior Program Officer
NANCY CAPUTO, Ocean Studies Board, Senior Project Assistant
DENISE GREENE, Ocean Studies Board, Senior Project Assistant

The work of this committee was overseen by the Ocean Studies Board and the Transportation Research Board of the National Research Council.

OCEAN STUDIES BOARD

KENNETH BRINK (*Chair*), Woods Hole Oceanographic Institution, Massachusetts
ARTHUR BAGGEROER, Massachusetts Institute of Technology, Cambridge
JAMES COLEMAN, Louisiana State University, Baton Rouge
CORTIS K. COOPER, Chevron Petroleum Technology Company, San Ramon, California
LARRY CROWDER, Duke University Marine Laboratory, Beaufort, North Carolina
G. BRENT DALRYMPLE, Oregon State University, Corvallis
EARL H. DOYLE, Shell Oil (ret.), Sugar Land, Texas
ROBERT DUCE, Texas A&M University, College Station
D. JAY GRIMES, University of Southern Mississippi, Ocean Springs
RAY HILBORN, University of Washington, Seattle
MIRIAM KASTNER, Scripps Institution of Oceanography, La Jolla, California
CINDY LEE, State University of New York, Stony Brook
ROGER LUKAS, University of Hawaii, Manoa
BONNIE MCCAY, Rutgers University, New Brunswick, New Jersey
RAM MOHAN, Blasland, Bouck & Lee, Inc., Annapolis, Maryland
SCOTT NIXON, University of Rhode Island, Narragansett
NANCY RABALAIS, Louisiana Universities Marine Consortium, Chauvin
WALTER SCHMIDT, Florida Geological Survey, Tallahassee
JON G. SUTINEN, University of Rhode Island, Kingston
NANCY TARGETT, University of Delaware, Lewes
PAUL TOBIN, Xtria, Chantilly, Virginia
JAMES YODER, University of Rhode Island, Narragansett

Staff

MORGAN GOPNIK, Director
SUSAN ROBERTS, Senior Program Officer
DAN WALKER, Senior Program Officer
JOANNE C. BINTZ, Program Officer
JENNIFER MERRILL, Program Officer
TERRY SCHAEFER, Program Officer
JOHN DANDELSKI, Research Associate
ROBIN MORRIS, Financial Officer
SHIREL SMITH, Office Manager
JODI BACHIM, Senior Project Assistant
NANCY CAPUTO, Senior Project Assistant
DENISE GREENE, Senior Project Assistant
DARLA KOENIG, Senior Project Assistant
JULIE PULLEY, Project Assistant

MARINE BOARD

RADOJE (ROD) VULOVIC (*Chair*), U.S. Ship Management, Inc., Charlotte, North Carolina
R. KEITH MICHEL (*Vice-Chair*), Herbert Engineering Corporation Alameda, California
PETER F. BONTADELLI, JR., PFB and Associates, Sacramento, California
BILIANA CICIN-SAIN, University of Delaware, Newark
BILLY L. EDGEPETER J. FINNERTY, Amercian Ocean Enterprises, Inc., Annapolis, Maryland
MARTHA R. GRABOWSKI, Rensselaer Polytechnic Institute, Cazenovia, New York
RODNEY GREGORY, PricewaterhouseCoopers LLP, Arlington, Virginia
I. BERNARD JACOBSON, Consultant, Long Island, New York
GERALDINE KNATZ, Port of Long Beach, Long Beach, California
SALLY ANN LENTZ, Ocean Advocates, Clarksville, Maryland
PHILIP LI-FAN LIU, Cornell University, Ithaca, New York
MALCOLM MACKINNON III, MSCL, Inc., Alexandria, Virginia
REGINALD E. MCKAMIE , Attorney, Houston, Texas
SPYROS P. PAVLOU, Environmental Risk Economics, URS Corporation, Seattle, Washington
CRAIG E. PHILIP, Ingram Barge Company, Nashville, Tennesee
EDWIN J. ROLAND, Elmer-Roland Maritime Consultants, Houston, Texas
E. G. WARD, Texas A & M University, College Station
DAVID J. WISCH, Texaco, Bellaire, Texas

Staff

JOEDY CAMBRIDGE, Marine Specialist
SUSAN GARBINI, Senior Program Officer

Acknowledgments

This report has been reviewed in draft form by individuals chosen for their diverse perspectives and technical expertise, in accordance with procedures approved by the National Research Council's Report Review Committee. The purpose of this independent review is to provide candid and critical comments that will assist the institution in making its published report as sound as possible and to ensure that the report meets institutional standards for objectivity, evidence, and responsiveness to the study charge. The review comments and draft manuscript remain confidential to protect the integrity of the deliberative process. We wish to thank the following individuals for their review of this report: Peter Bontadelli (PFB and Associates), Michel Boufadel (Temple University), David Fritz (BP/AMOCO), Stephen Monosmith (Stanford University), Jeffrey Short (National Marine Fisheries Service), Christopher Reddy (Woods Hole Oceanographic Institution), Sylvia Talmage (Oak Ridge National Laboratory).

Although the reviewers listed above have provided many constructive comments and suggestions, they were not asked to endorse the conclusions or recommendations nor did they see the final draft of the report before its release. The review of this report was overseen by Andrew Solow, Woods Hole Oceanographic Institution. Appointed by the National Research Council, he was responsible for making certain that an independent examination of this report was carried out in accordance with institutional procedures and that all review comments were carefully considered. Responsibility for the final content of this report rests entirely with the authoring committee and the institution.

Preface

Spills of Emulsified Fuels: Risks and Response is part of an evolving body of work conducted by the National Research Council (NRC) to help inform debate and decision-making regarding the ecological consequences of releases associated with the widespread use of fossil fuels. Like earlier NRC reports, it attempts to understand the chemical, physical, and biological behavior of a complex mix of compounds that make up various petroleum hydrocarbon-based fuels. The specific risk factors presented by emulsified fuels are difficult to characterize, mainly because there have been no spills of emulsified fuels to date, and thus there is little practical experience with these products.

The Committee on Spills of Emulsified Fuels: Risks and Response faced some special challenges in the conduct of this study. First, it had to evaluate a new product with very little real-world data for validation of the assumptions used in models and risk assessments. In addition, the committee had to evaluate a new formulation as of 1998, when much of the early research results were for the previous formulation. It became necessary to decide which results could be applied to the new formulation and which results had to be set aside. Second, Orimulsion® is a complex, multicomponent fuel that, when spilled, behaves very differently than other known types of oil. One of the first activities of the committee was to create its own conceptual models of how this fuel would behave when spilled in different combinations of water salinity, rates of diffusion and dilution, and current speeds. During this process, committee members developed a better understanding of how this product behaves when spilled and the potential for impacts on sensitive resources. These conceptual models became the basis for discussions in each of the chapters on fate and behavior, effects, and response.

Another challenge was how to evaluate the impacts of a spill on top of background levels of widely used chemicals (namely, a specific group of surfactants) that are beginning to be considered as chronic contaminants of concern. Spills by their nature are unpredictable and episodic. How does one evaluate the potential impacts of such events in light of the background of other sources of environmental stressors? There has been a growing recognition of the potential for impacts from surfactants that are widely used in household products as well as industrial applications. The committee had to evaluate potential impacts from spills within the framework of this growing, yet undefined, level of concern for the chronic environmental impacts of this group of surfactants.

Despite these hurdles and the tight time frame, what follows is, in my opinion, a very credible review of the work completed to date and an evenhanded discussion of the remaining questions. As with all such efforts, only the hard work of the committee members and the NRC staff made the report possible. I would like to take this opportunity to recognize their contributions.

> Jacqueline Michel
> *Chair*, Committee on Spills of Emulsified Fuels:
> Risks and Response

Contents

EXECUTIVE SUMMARY		1
1	INTRODUCTION AND OVERVIEW	7
2	BEHAVIOR AND FATES: SUMMARY AND EVALUATION OF AVAILABLE INFORMATION	14
3	ECOLOGICAL EFFECTS: SUMMARY AND EVALUATION OF AVAILABLE INFORMATION	44
4	EFFICACY OF RESPONSE: SUMMARY AND EVALUATION OF AVAILABLE INFORMATION	66
REFERENCES		81
APPENDIXES		
A	COMMITTEE BIOGRAPHIES	89
B	LITERATURE REVIEWED BY THE COMMITTEE	92
C	ACRONYMS	104

Executive Summary

As the demand for electricity grows in the United States, power generators are looking for alternative fuels and stable prices. Among the alternatives being considered are a group of multi-component fuels referred to as emulsified fuels. One such fuel, is known as Orimulsion® (referred to simply as Orimulsion throughout the report), produced by Bitúmenes Orinoco, S.A. (BITOR), a subsidiary of the Venezuelan national oil company Petroleos de Venezuela, S.A. (PDVSA). Orimulsion is a multi-component fuel composed of roughly bitumen (70 percent), fresh water (30 percent), and two additives (a surfactant and a stabilizer, <0.2 percent). Bitumen is an asphaltic or heavy tar-like mixture of hydrocarbons with from 10 to more than 1,000 carbon atoms that occurs naturally or is obtained as highly viscous residues after refining (distillation) of crude oils to remove most lighter-molecular-weight components. The natural bitumen used in Orimulsion comes from the Orinoco belt in Venezuela, which has one of the largest reserves of petroleum in the world. For information on petroleum formation and the natural processes that affect petroleum the NRC report *"Oil in the Sea"* (1985) is recommended.

The extensive reserves and limited market for bitumen allow BITOR to offer long-term contracts at fixed rates significantly lower than many other fuels. Orinoco bitumen is highly weathered and too viscous to flow; thus, BITOR developed many innovative methods to extract the bitumen and produce a fuel that can be used in power generation. A diluent (kerosene) is added at the wellhead to help move the bitumen from the production fields to the Orimulsion manufacturing facility. The diluent is completely removed, and an emulsion is formed by mixing the bitumen at high energy with a surfactant, a stabilizer and fresh water from the Morichal River. The Morichal River is a relatively pristine river with no human or industrial activities nearby. The water is filtered and chlorinated to remove any bacteria. In 1998, BITOR modified the formulation, and the new product is marketed as Orimulsion-400.

In 1994, BITOR implemented special precautions in spill prevention and safety measures. These measures include the use of only double-hulled vessels both at sea and in rivers, and requirements for special procedures and equipment on rivers. The vessels are prescreened through vetting procedures with special criteria.

During the transport and storage of fuels including Orimulsion, there is always the risk of accidental spills. As with other liquid petroleum products, transport of this fuel requires development of an emergency response plan. However, Orimulsion has unique properties that make it behave very differently from conventional petroleum products. First, the density of the emulsified product is intermediate between fresh water and salt water. Therefore, Orimulsion will float, sink, or suspend in the water column, depending on the water density. Second, the emulsion breaks when diluted sufficiently with water. Therefore, following a spill the components separate into the bitumen droplets (with a coating of surfactant) and the freshwater component that contains the dissolved polynuclear aromatic hydrocarbons (PAH), surfactant, and stabilizer. In this case, the fate and possible effects of each component must be evaluated separately.

These unique properties make the use of standard spill response techniques less effective in tracking, containment, recovery, and cleanup of Orimulsion spills. BITOR recognized these problems and has spent significant time and resources in conducting studies of the environmental behavior, fate, and effects of Orimulsion spills and in developing specialized techniques to support spill response plans. Because there have not been any Orimulsion spills, regulators are faced with having to evaluate these studies and technologies without the practical experience that comes from an actual response to a spill. Therefore, the U.S. Environmental Protection Agency (EPA) and U.S. Coast Guard (USCG) requested the NRC to undertake a fast-track study to assess the validity and usefulness of the current work on a variety of emulsified fuels. However, after examining the available literature, the committee concluded that adequate literature was available only for Orimulsion. Thus, this report deals almost exclusively with Orimulsion. BITOR, the company that manufactures Orimulsion, funded many of the studies on the behavior of emulsified fuels cited in this report. Except where specifically discussed, the committee found the studies described in the available literature to be well executed and documented, and to have followed established laboratory practices. Consequently, the committee found no reason to question the validity of the analyses reported. **As noted throughout the remaining chapters, the committee did identify areas in which study** design and the resulting interpretations should be reexamined and some underlying assumptions reevaluated.

UNDERSTANDING THE PROBLEM

To describe the potential environmental effects of Orimulsion spills or identify information needed to support spill response planning, it is necessary to understand how the product behaves when released into the environment. The breakdown of the emulsion when diluted with water releases a cloud of bitumen droplets. In fresh water, the surfactants attached to the bitumen droplets retard coalescence of the bitumen; however, an increase in water salinity to about 6 parts per thousand renders the surfactants ineffective. Breakdown of the surfactant permits the bitumen droplets to agglomerate or coalesce (although the conditions of this process are not well understood) and form larger droplets, which increases the likelihood of them floating in salt water. Re-floated bitumen will form sticky tarballs and patties that behave similarly to tarballs from other types of heavy fuel spills. The ultimate fate of the bitumen will depend on the spill location and conditions, but those droplets that do not re-float will ultimately be degraded and deposited on the bottom. The bitumen droplets do not adhere to suspended particulate material in fresh water, because traces of surfactant adhering to the bitumen appear to be effective even after dilution. However, in brackish or full strength seawater, bitumen will likely adhere to suspended particulate material (SPM) that can be transported to the bottom or onto the shoreline.

When the emulsion breaks, the water component of Orimulsion that is released contains PAH dissolved from the bitumen droplets as well as the two additives: a water-soluble nonionic surfactant (a mixture of widely used alcohol ethoxylates [AE]) and an emulsion stabilizer (monoethanolamine). The additives are biodegradable and, because of their high water solubilities, are unlikely to bioaccumulate. However, the fate and toxicity of biodegradation products have not been characterized completely.

As with spills of any material, site-specific environmental factors to a large degree control the fate and effect of spills of emulsified fuels such as Orimulsion. In an effort to articulate general concerns that vary with the changes in environmental conditions, six different scenarios for Orimulsion spills in open and closed water bodies, at different salinities and with varying degrees of turbulence, were developed by the committee. It is clearly beyond the scope of this study to develop site-specific, quantitative discussions of the fates and effects of spills of emulsified fuels that could occur in the future. Therefore, the committee developed qualitative descriptions of the expected fates of the dispersed bitumen droplets and dissolved constituents as a function of time for a range of environmental settings. These scenarios provide a summary of general conclusions regarding spill behavior and a basis for later assessment of the environmental impacts of spills and the efficacy of spill response strategies.

OVERARCHING GAPS IN KNOWLEDGE

The bitumen used to prepare Orimulsion is highly weathered (degraded). Thus, Orimulsion has very low concentrations of volatile organic components, and total benzene, toluene, ethylbenzene, and xylene (BTEX) concentrations are an order of magnitude lower than those observed in a typical No. 6 fuel oil. More importantly, total PAH (the major source of toxicity in oils) concentrations in Orimulsion are very low and up to one order of magnitude below those typically found in No. 6 fuel oil. The concentration of PAH in the water phase of Orimulsion and the receiving water body after a spill is largely controlled by the rates of absorption and desorption of PAH onto bitumen droplets in the neat fuel or bitumen droplets and suspended particulate matter in the receiving waters. These processes can be described by equilibrium partitioning theory, which predicts that additional PAH should desorb from the bitumen droplets (especially given their high surface-area-to-volume ratios) when the water phase is sufficiently diluted. Studies of Orimulsion using standard techniques for describing equilibrium partitioning theory indicate that the maximum theoretical concentration that could be seen by exposed organisms cannot exceed the effective initial concentration of dissolved PAH in the neat fuel (15-30 parts per billion). **Although reasonable, these theoretical studies have to be verified independently using analytical laboratory techniques at a variety of dilution factors, as part of a specifically designed study. Furthermore, comparison of Orimulsion to other fuels (such as No. 6 fuel oil) would necessitate that similar studies be performed on the fuels of interest.**

Because there are significant limitations in containment and recovery of spilled Orimulsion, past research has been concentrated on spill prevention and on understanding the behavior and effects of Orimulsion spills. One of the most significant data gaps in understanding the behavior and fate of spilled Orimulsion is the degree of coalescence of the bitumen droplets that are expected to form early in a salt water spill when the concentration is high. Currently, there are insufficient data to accurately predict the percentage of a spill that will surface or sink due to coalescence. **Further research is needed to quantify the competing processes of coalescence and dispersion for different spill volumes, release rates, and turbulent energy.**

In the absence of actual incidents, models must be used to predict the behavior of Orimulsion spills. At present, current models have not been validated, yet spill response planners rely on these models to develop response strategies. To support better site-specific response plans and identify the types of equipment that should be available, improved models have to be developed. **Additional studies using a wider variety of particulate types are needed to address** uncertainties in how bitumen interacts with suspended sediments, particularly those with high organic content. Furthermore a validated model predicting spill behavior should be completed and model output should be verified for site-specific application.

POTENTIAL ENVIRONMENTAL EFFECTS

The composition and behavior of Orimulsion directly influences the environmental effects of the spilled product. For spills in fresh and brackish water, where none of the bitumen will be on the surface, direct impacts on shorelines and animals that use the water surface (birds, mammals) will be greatly reduced (compared to floating oil spills). In salt water, the potential for impacts on these resources will be a function of how much of the bitumen resurfaces. Conversely, compared to floating oils, potential impacts on benthic and water column resources from spills of Orimulsion may be increased. Because the bitumen is so highly weathered and the effective concentration of PAH in the water phase is also low, the amount of toxic PAH biologically available is also low. This suggests that acute impacts from PAH toxicity will be negligible except in settings where the plume dilutes very slowly. **However, bioassay tests completed to date have not been supported by adequate and complementary chemical analyses to determine the concentration of exposure to PAH or the surfactant. Further studies are needed to better document the role of PAH partitioning in Orimulsion bioassay tests or in comparison tests involving other fuels.**

Most of the studies conducted to date to evaluate environmental effects from possible spills have focused on water column exposures of marine fish and invertebrates. There are no studies of the effect of a spill plume on corals or the direct exposure of the undiluted fuel on vascular plants, and there are only limited studies of the effect on other nonplanktonic primary producers. Because the sediments are the ultimate sink for bitumen droplets, studies are needed to evaluate the bioavailability of bitumen associated with sediments and the effects on plants and animals from long-term exposure to bitumen-contaminated sediments. Thus ingestion or contact with these droplets by benthic organisms represents a viable pathway of exposure. **There is insufficient information on the bioavailability to and potential impacts of PAH on benthic animals from long-term exposure to bitumen-contaminated sediments. Additional study is needed to better understand the potential risk faced by these organisms.**

The water fraction of Orimulsion contains two additives. The stabilizer, monoethanolamine (MEA) has low toxicity and poses no apparent risk to aquatic resources. The surfactant mixture, composed of alcohol ethoxylates (AE), is widely used in household and industrial products; thus, potential impacts have to be considered in light of the background concentrations of these chemicals and their intermediary decay products in rivers and streams that receive sewage discharges. Under most conditions, concentrations are expected to decrease quickly below no-effect levels for the parent compounds. There is uncertainty, however, about the toxicity and degradation rates of intermediate breakdown products. To place the potential impact of the release of these surfactants (whether through a spill of Orimulsion or from other existing sources through the wastewater stream) in context given their widespread use, more information on their background concentrations and behavior in the environment is needed. **Federal and state agencies should consider developing information on the ambient concentration of surfactants and their degradation products.**

EFFECTIVENESS OF SPILL RESPONSE STRATEGIES

Because Orimulsion is pre-dispersed, the initial response strategy will be to track the suspended plume and monitor for any surfaced tarballs. The tarry nature of the surfaced bitumen requires specialized skimming, pumping, and handling systems, and such systems have been prototyped and tested under limited conditions. For spills in fresh and brackish water, most response options are not applicable because very little of the spill is expected to float initially or re-float over time. Cleanup will likely consist of removal of bitumen droplets that have accumulated on the bottom in low-flow areas by dredge or vacuum systems. **Many of the proposed methods should be further refined to improve their effectiveness, and responders should become familiar with the effective use of these methods prior to a spill emergency.**

1
Introduction and Overview

Over the last decade, the United States has experienced a growing demand for energy, a demand that is expected to increase at a remarkable rate. During this same time, environmental concerns have continued to drive exploration into a variety of options to minimize adverse environmental impacts associated with extraction, transportation, and consumption of fossil fuels. Although electric power can be generated using hydropower, nuclear energy and the burning of fossil fuels (including coal, natural gas, and liquid petroleum), there has been limited historical use of oil to produce electricity (National Energy Policy Development Group, 2001). Recent work completed by the Energy Information Administration (2001) of the Department of Energy indicates that electricity generation fueled by natural gas and coal will increase through 2020 to meet growing demand for electricity, and offset the projected retirement of existing nuclear units (Figure 1.1). Projections for power generation from natural gas, coal, and nuclear power are expected to increase as a result of higher projected electricity demand and the improved operating costs and performance of nuclear plants. The use of renewable energy technologies for electricity generation is projected to grow slowly because of the relatively low costs of fossil-fired generation and because electricity restructuring favors less capital-intensive natural gas technologies over coal and baseload renewables. As also shown in Figure 1.1, the firing of liquid petroleum products, such as No. 6 fuel oil, accounts for a relatively small and declining fraction of electricity generation (Energy Information Administration, 2000). Despite this overall trend, electric generators have been increasing their use of Group V fuels (often referred to as low API [American Petroleum Institute] gravity oil, or LAPIO [Low API oil]) because of the rela-

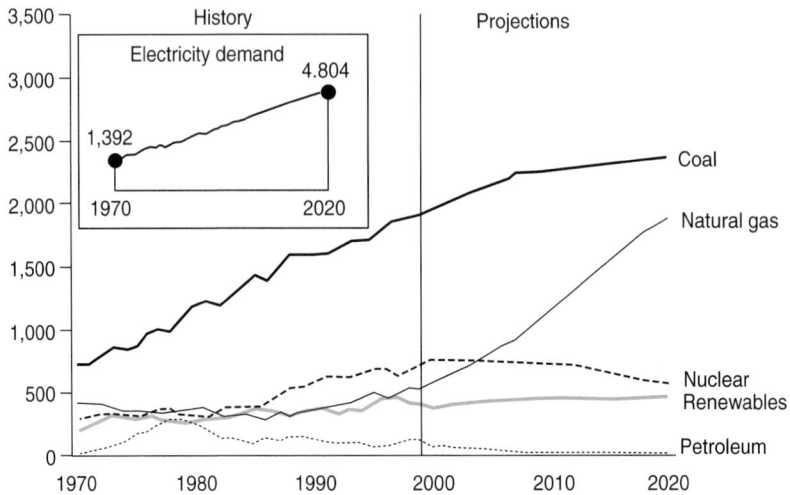

FIGURE 1.1 Electricity generation by fuel, 1970-2020 (billion-kilowatt hours) (Energy Information Administration, 2000).

tively low cost and high Btu (British thermal unit) values (Michel et al., 1995; National Research Council, 1999). Included in this group is a special class of fuels termed emulsified fuels.

UNDERSTANDING EMULSIFIED FUELS

Emulsified fuels are multicomponent fuels analogous to water-in-oil emulsions. These fuels may possess combustion properties considerably different (and in some instances more favorable) than those of the base fuels from which they are formed due to the presence of different compounds in the reaction environment and the physical changes in atomization behavior (Miller and Srivastava, 2000). Emulsified fuels include three major variants, coal-water slurries, water-in-oil emulsions, and bitumen-water emulsions.

Coal-based multicomponent fuels were developed primarily in the 1970s and 1980s as an alternative to fuel oils. However, when the price differential between crude oil and coal decreases, interest in coal-based multicomponent fuels also decreases (Miller and Srivastava, 2000). Conversely, the volatility of crude oil prices (especially when compared to the price of coal, which has historically been stable) makes conversion to liquid petroleum-based fuels less appealing in the long run. Similarly, water-in-oil emulsions tend to vary in price more radically than coal (though less so than crude oil[1]). These emulsified fuels are designed to

---

[1] In general, there is very little production of emulsified fuels for use in the United States. Annual production and sales of emulsified fuels seem to be in the range of about 190,000 to 240,000 barrels

improve the combustion properties of the petroleum component (usually a distillate; Miller and Srivastava, 2000). By far the best known and studied emulsified fuel[2] is a bitumen-water emulsion named Orimulsion produced by Bitúmenes Orinoco, S.A. (BITOR), a subsidiary of the Venezuelan national oil company Petroleos de Venezuela, S.A. (PDVSA). Unlike oil, the price of bitumen is extremely stable and predictable. The large reserves and limited market for bitumen have allowed BITOR to offer long-term contracts at attractive rates. Thus, Orimulsion is seen as being cost-competitive with other power generation fuels leading to greater interest in its use worldwide.

The use of Orimulsion requires similar infrastructure to that used to burn liquid petroleum for power generation. Power generation plants that currently use No. 6 fuel oil are the most easily converted to the use of Orimulsion. Therefore the most common comparison for risk assessment of spills of Orimulsion in the literature is with No. 6 fuel oil. The comparison in this report therefore reflects the most common potential uses of Orimulsion.

Orimulsion is composed of bitumen (~70 percent), fresh water (~30 percent), and surfactant and stabilizer (<0.2 percent). As its name implies, the bitumen is produced from the Orinoco belt in Venezuela, one of the largest reserves of petroleum in the world (estimated to be 300 billion barrels). Orinoco bitumen is highly weathered and extremely viscous (viscosity > 10,000 cP at 30°C; for comparison the viscosity of water is 1 cP; honey, 10,000 cP; and molasses,

---

per year. There has been more production and use in the past, but many customers seem to have switched to natural gas. There is hope on the part of fuel and surfactant suppliers that increased natural gas prices will help to shift interest back to emulsified fuels. At this time that does not seem to be happening to a major degree, but one supplier did note that its sales increased by about 40 percent from 1999 to 2000 due to gas price increases. All current sales discussed by the companies contacted were to industrial customers as opposed to utilities. All fuel sales were on the East Coast, either in the Northeast or in North Carolina. Contacts at these companies were not aware of any emulsified fuel suppliers other than those discussed here.

[2]There is no category for emulsified fuels in the *Thomas Register*, and a search of various combinations of "emulsion, oil, fuel, petroleum" does not identify any company that produces emulsified fuel oils. However, the Environmental Protection Agency has conducted tests of fuels from two companies that market emulsified fuel oils, Industrial Fuel Company of Hickory, N.C., and Clean Fuels Technology (CFT) of Reno, Nev. (formerly A-55 Clean Fuels, Ltd.).

CFT's business is almost entirely emulsified fuels, with groups that market both to large stationary plants (industrial and utility boilers and gas turbines) and to transportation users (diesel engines). The boiler fuel is typically an emulsified heavy (No. 6) fuel oil, while the turbine and diesel fuels are lighter fuel oils (No. 2, diesel fuel) emulsified with water. CFT has licensed its technologies to different suppliers. Lubrizol markets the diesel fuel as "PuriNOx" fuel nationally, and Global Petroleum is the only current company to market emulsified heavy fuel oil using the CFT technology. The CFT approach is to emulsify the fuel on-site where possible. According to CFT, its technologies can be used to emulsify any heavy petroleum product into a fuel. For some applications (asphalt or refinery bottoms), the emulsification may be required prior to shipment. There are no current applications of this type in the United States.

100,000 cP). A diluent (kerosene) is added at the wellhead to reduce viscosity and to ease pumping of the bitumen from the production fields to the manufacturing plant. The diluent is separated from bitumen by flash heating. Water from the Morichal River, an alcohol ethoxylate surfactant (AE), and a monoethanolamine stabilizer (MEA) are added to the bitumen and oil components and mixed under high energy to form an emulsion with a viscosity of about 300 cP (similar to a light crude oil). Orimulsion is then readily pumped via pipeline to a coastal terminal for loading onto ships. The Orimulsion formulation was modified in 1998, when BITOR changed the surfactant from nonylphenol ethoxylates, because of concern that these compounds were potential endocrine disrupters and the degradation metabolites were more toxic and persistent than the parent compounds. In addition, magnesium is no longer added. The previous formulation is referred to as Orimulsion-100, and the new formulation is marketed as Orimulsion-400.

SPILL RESPONSE ISSUES

Following the Oil Pollution Act of 1990, oil spill response has focused on improving the capability to contain and recover spilled oil. Response plans for facilities and vessels have to include sufficient resources that can be deployed effectively to recover specified amounts of oil within a set time (usually 72 hours). Area contingency plans identify priority protection areas and develop site-specific plans to protect the most sensitive resources. Even with this emphasis on preparedness, it is widely accepted in the spill response community that on-water recovery rates of 20 percent of the spilled oil are rarely exceeded. Nearly all of the response plans, strategies, and equipment are based on the assumption that the oil will float. The 1999 National Research Council report *Spills of Nonfloating Oils,* found that planning for spills of non-floating oil was inadequate, lacking in equipment, response plans, and cleanup methods. To support its own spill response planning, BITOR has implemented special precautions in spill prevention and safety measures, starting in 1994. These measures include the use of only double-hulled vessels both at sea and in rivers and requirements for special procedures and equipment on rivers. The vessels are prescreened through vetting procedures with special criteria.

The unique properties and behavior of emulsified fuels suggest that some scrutiny be given to how the response to spills of this type may differ from those of more traditional, floating liquid petroleum products or crude oil. The density of Orimulsion is about 1.01 g/ml at 15°C, which is greater than that of fresh water (=1.00 g/ml) but less than seawater (=1.025 g/ml). The API gravity ranges between 7.8 and 9.3; thus, it is characterized as a Group V oil under U.S. Coast Guard regulations. Spills of Orimulsion pose special issues for spill response because the fuel is essentially predispersed. When spilled into water, the emulsion breaks down, releasing the bitumen droplets. Depending on the salinity and currents in the receiving water, these particles can either float, sink, or be kept in suspension. In areas with high concentrations of bitumen, the droplets can recoalesce and rise to the surface, forming tarry slicks. The eventual fate of the bitumen droplets varies according to the spill conditions, although very little recovery of the bitumen particles is likely.

ORIGIN OF STUDY

Congress, through the FY 1997 appropriations bill, directed the U.S. EPA to initiate research into the qualities and characteristics of Orimulsion and its potential environmental impact. Although not currently used by utilities in the United States, Orimulsion is used in Canada, Japan, China, Italy, and other nations. In response to congressional direction, the EPA studied combustion emissions and control and characterized the risk of Orimulsion use in power plants, including potential impacts of spills. The results of the EPA studies are discussed at length in a recent report to Congress (Miller et al., 2001).

Proposals to import Orimulsion into the United States have generated environmental concerns due, in part, to limited information regarding the impact of potential spills. Because of the location of utilities along major rivers and lakes, coupled with the low cost of moving fuel oils by barge, these environments could be exposed to risk in the event of a significant spill. Existing research into the impact and cleanup of spills of emulsified bitumen, however, focuses largely on the marine (saltwater) environment. In an effort to further assess the validity and usefulness of existing literature pertaining to possible spills of Orimulsion, the U.S. EPA and the U.S. Coast Guard requested the National Research Council to undertake a fast-track study to review and evaluate the work on Orimulsion completed to date. Specifically, the sponsors asked that the study "describe the potential environmental impacts of transport-related spills of emulsified fuels, with emphasis on emulsified bitumen, in marine and fresh waters and specify the information needed to evaluate and respond to these risks" (Box 1.1). Furthermore, the committee was asked to "consider relevant literature on transport-related spills of other emulsified fuels, such as emulsified petroleum products. The adequacy of the available research will be assessed and requirements for future research, if any, will be described" and to "determine if specific improvements are needed for response and clean-up of spills of emulsified fuels in both marine and fresh water."

The committee's evaluation focused on what information was needed to improve oil spill contingency planning and response. Like the NRC (1999) report on the risk and response for non-floating oil spills, the committee looked at the issues from spills of emulsified fuels from a responder's perspective and the

> **Box 1.1**
>
> **Statement of Task**
>
> This study will describe the potential environmental impacts of transport-related spills of emulsified fuels, with emphasis on emulsified bitumen, in marine and fresh waters and specify the information needed to evaluate and respond to these risks. An overview and analysis of the literature, principally available for Orimulsion (a trademarked product of BITOR), on the potential consequences of such spills will be provided.
>
> The study will also consider relevant literature on transport-related spills of other emulsified fuels, such as emulsified petroleum products. The adequacy of the available research will be assessed and requirements for future research, if any, will be described. The final report will determine if specific improvements are needed for response and clean-up of spills of emulsified fuels in both marine and fresh water.

standards of preparedness and research that have been applied to oil spill response over the last ten years.

The list of literature reviewed during this study, consisting of more than 300 publications, is included in the Reference section and in Appendix B. Most of the research about the behavior of emulsified fuels in water has been funded, or partly funded, by BITOR, and deals only with Orimulsion; work in this area, in partnership with the U.S. Coast Guard and Environment Canada, continues. Additional information on the fate and behavior of polycyclic aromatic hydrocarbons and the surfactants in Orimulsion was also reviewed. The committee found that, of the emulsified fuels, adequate literature to complete its task was available only for Orimulsion. As is common with research directed at a specific product or proprietary process, the overwhelming majority of the available studies were funded by BITOR, the producers of Oriumulsion (or entities interested in using it). This raised some question of the independence and hence the validity of the studies completed. Except where specifically discussed in this report, the committee found the studies to be well executed and documented and to have followed established laboratory practices. Consequently, the committee found no reason to question the validity of the analyses reported. **As noted throughout the remaining chapters, the committee did identify areas in which study design and the resulting interpretations should be reexamined and some underlying assumptions reevaluated.**

Many of the general technical issues raised about Orimulsion during this study have analogues with spills of other emulsified fuels or spills of crude oil or liquid petroleum products. Where possible, the implications of these issues are discussed in terms of general spill effects or responses. **However, since much of**

the material in the following chapters deals almost exclusively with Orimulsion, careful consideration should be used before any specific conclusions are extended to other emulsified fuels.

Based on available literature, the committee developed a number of findings and recommendations that are presented in the following chapters. Environmental factors, such as salinity, affect the behavior of Orimulsion; thus, spills in different environmental settings may provide unique challenges or risks. Although it is beyond the scope of this study to develop quantitative site-specific discussions of the fates and effects of spills of emulsified fuels, six qualitative spill scenarios are presented. Chapter 2 summarizes and evaluates the available literature on the physical and chemical characteristics of Orimulsion as they affect the predicted behavior and fate of the various components once Orimulsion is spilled into water and on land. Similarly, Chapter 3 summarizes, in general terms, various aspects of the potential effects of Orimulsion spills in various environments. Chapter 4 summarizes and evaluates the available proposed response strategies and equipment for responding to Orimulsion spills.

2

Behavior and Fates: Summary and Evaluation of Available Information

BEHAVIOR AND FATE

Physical Characterization

Much of the scientific research on the fate of emulsified fuels has been focused on a particular commercial product called Orimulsion manufactured in Venezuela. The following properties specifically refer to Orimulsion-400, which replaced an earlier version of the product called Orimulsion-100. Orimulsion is a mix of approximately 70 percent natural bitumen (pitch) and 30 percent fresh water, with additives to maintain the emulsion. The additives are 1,100 parts per million (0.11 percent) MEA, an emulsion stabilizer, and 1,350 ppm (0.135 percent) AE, a water-soluble nonionic surfactant (Golder and Associates, 2001) that retards the coalescence of bitumen droplets. The designations 100 and 400 refer to the type of nonionic surfactant used in the formulation. (Much of the existing literature discusses studies of Orimulsion-100 thus this report makes the distinction as needed.)

With a specific gravity greater than 1.0, Orimulsion is classified for regulatory purposes as a Group V oil (Stout, 1999). The density of the product is intermediate between fresh and salt water (Figure 2.1). It is comparable in pour point[1] and viscosity to an industrial fuel oil. The dynamic viscosity of Orimulsion varies with temperature but is in general lower than that of many Group V oils.

---

[1] The lowest temperature at which a substance, such as oil, will flow under specified conditions.

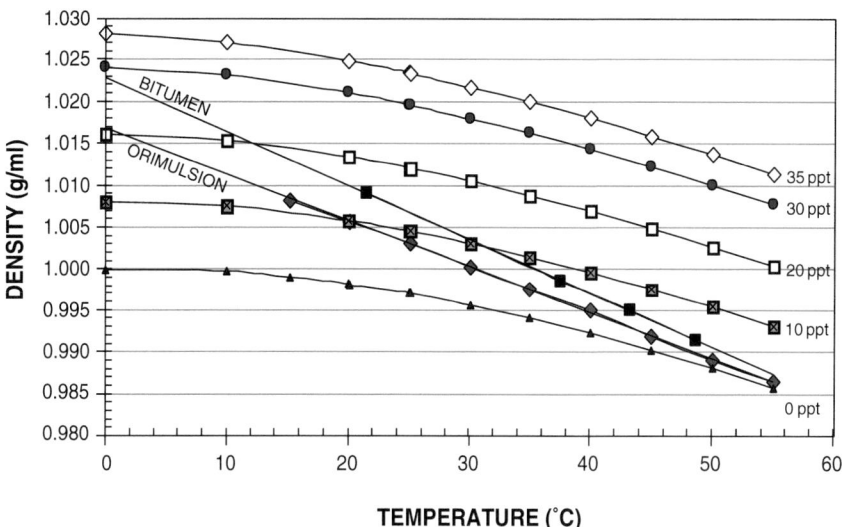

FIGURE 2.1 Density-temperature relationship for Orimulsion-400, bitumen, and different salinity waters. SOURCE: Reprinted from PDVSA–Intevep,1998. Permission granted by Bitor America Corporation.

The flash point is > 95°C, and thus Orimulsion is classified as nonflammable by the National Fire Protection Association (NFPA).

Pure bitumen has a high viscosity, on the order of a million centipoise at typical ambient water temperatures. The density and the pour point are higher than those of the emulsified Orimulsion product (Figure 2.1). Jokuty et al. (1995) list a pour point of 38°C for bitumen versus a pour point of 3°C for fresh Orimulsion. In completely quiescent conditions, dispersed bitumen droplets would be expected to sink or to be neutrally buoyant in water at 15°C and salinity less than 10-15 parts per thousand (ppt)[2] (Crosbie and Lewis, 1998a), but they would be expected to float in more saline water.

The average bitumen droplet size in fresh Orimulsion is around 20 μm (Ostazeski et al., 1998a, 1998b). Although the droplet size distribution (Figure 2.2) is somewhat bimodal (Stout, 1999), almost all of the droplets range in size from 1 to 80 μm. Exposure to even low-salinity water (>5-7 psi (Practical Salinity Units) collapses the surfactant in Orimulsion (Brown et al., 1995; Crosbie and Lewis, 1998b); releasing the bitumen from the emulsion where it can form

---

[2]The documents examined report concentrations in metric units such as mg/L or parts per million. This difference reflects differences in common use between scientific practice within the community. This report may use either or both, depending on use in the literature reviewed. For the concentrations and conditions discussed in this report, the conversion between systems is unwarranted.

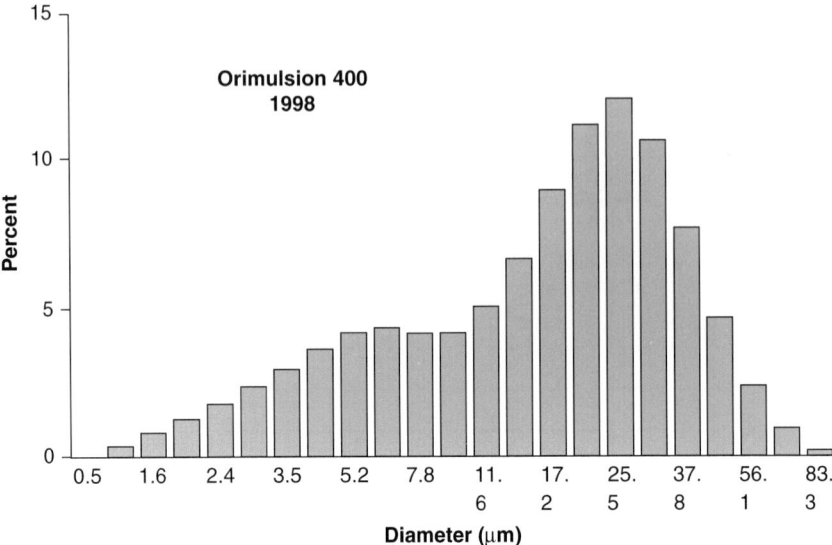

FIGURE 2.2 Histogram showing average bitumen droplet size in fresh Orimulsion-400. SOURCE: Stout, 1999.

droplets. If the droplets collide, the bitumen may coalesce and form larger droplets, which (based on Stokes equation, Figure 2.3), increases the likelihood that the droplet will either surface or sink, depending on its relative buoyancy compared to the surrounding water.

Because Orimulsion is not a homogeneous fluid, its properties may change significantly during the course of an accidental release. According to Febres et al. (1995), at a concentration greater than 20,000 ppm the product retains its emulsion properties, while at a concentration less than 10,000 ppm the material is expected to behave like dispersed bitumen droplets. Such high concentrations would exist only for an open-water spill within the immediate vicinity of the spill source and for a very short time. The exception would be a spill scenario in which mixing with the ambient water and subsequent dilution of the product were restricted. For the most part, predicting Orimulsion spill behavior becomes a matter of predicting the behavior of the dispersed bitumen cloud.

Chemical Characterization

The Cerro Negro bitumen used to produce Orimulsion comes from the Orinoco Belt in the Eastern Venezuelan Basin (Bitumenes Orinoco). This bitumen is highly weathered (degraded) in nature and consists primarily of high-molecular-weight, multi-ring aromatic hydrocarbons and resins that account for

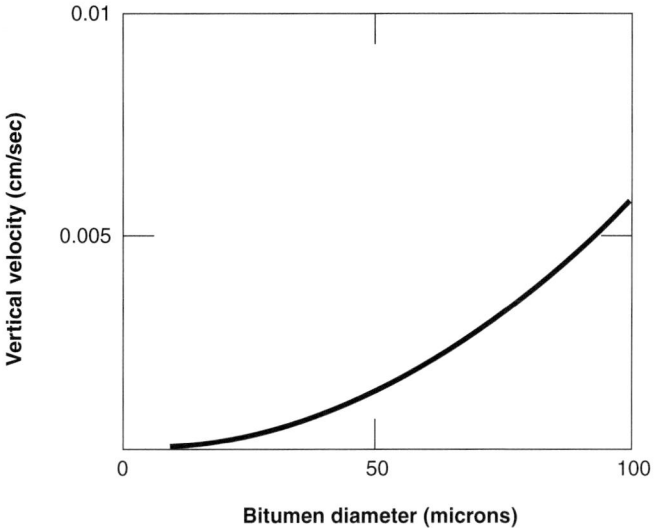

FIGURE 2.3 Re-float velocity of bitumen droplets (assumed to be spherical) by Stokes equation in seawater (35 ppt) at 15°C.

63 to 69 percent of the fuel (Ostazeski et al., 1998a, 1998b; Brown et al., 1995; Jokuty et al., 1999; Stout, 1999). Figure 2.4 presents flame ionization detector gas chromatographic profiles. These profiles show neat Orimulsion (sample 3), an oil-in-water dispersion of 18,250 mg/L Orimulsion (dissolved phase) filtered through a 1-µm (micron) membrane (sample 4), and an oil-in-water dispersion of 5,475 mg/L Orimulsion that has not been filtered (sample 8), (Wang and Fingas, 1996). The neat Orimulsion is characterized by a bimodal unresolved complex mixture (Fig. 2.4A). Very few of the individual constituents present in the Orimulsion are partitioned into the dissolved phase (Fig. 2.4B). The unfiltered sample shows the total petroleum hydrocarbons clearly associated with the bitumen droplets as evidenced by the pattern which is almost identical to the whole Orimulsion pattern (Fig. 2.4C).

Orimulsion has very low concentrations of BTEX that are at least an order of magnitude lower than in the typical No. 6 fuel oil it is likely to replace (Table 2.1). Individual and total polynuclear aromatic hydrocarbon (PAH) concentrations in Orimulsion are up to one order of magnitude below those typically found in crude oils and refined products (Table 2.2). Orimulsion is relatively high in sulfur, nickel, and vanadium (Table 2.1), with the latter two constituents tied up as metalloporphyrins, which make them biologically unavailable.

The additives used in the production of Orimulsion-400 are a water-soluble nonionic surfactant (a narrow distillate cut of widely used AE) and an emulsion

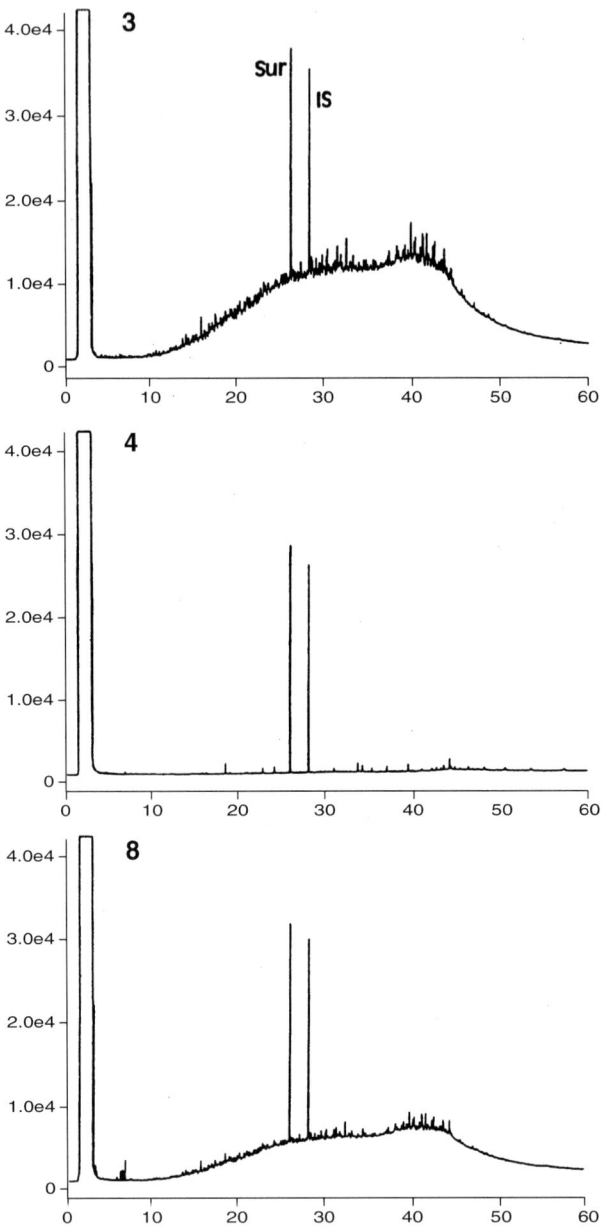

FIGURE 2.4 Flame ionization detector gas chromatographic profiles for neat Orimulsion (sample 3), an oil-in-water dispersion filtered through a 1-µm membrane (sample 4), and an oil-in-water dispersion that has not been filtered (sample 8). SUR and IS represent surrogate and internal standards, respectively. SOURCE: Wang and Fingas, 1996. Copyright 1996 American Chemical Society.

TABLE 2.1 Physical Properties and Chemical Composition of Orimulsion-400 and No. 6 Fuel Oil

Parameter	Orimulsion-400	No. 6 fuel oil
Density (g/ml at 15°C)	1.01-1.02	0.94-1.02
Pour point (°C)	0-3	−17-36
Viscosity (cP)	200-350 (at 30 °C)	325-47,000 (at 15 °C)
Mean particle size (μm)	14-20	Not applicable
Sulfur content (wt%)	2.85	0.7-3.0
Hydrocarbon groups (wt%)		
Saturated	14	11
Aromatic	47	55
Resins	22	20
Asphaltenes	17	14
Total BTEX (ppm)	36	464
Benzene	0	17
Toluene	4	100
Ethylbenzene	19	47
Xylene	13	300
Total PAH (μg/g bitumen)	3,040	317,627
Naphthalenes (C_0-C_4)	474	46,600-106,000
Phenanthrenes (C_0-C_4)	854	50,200-113,000
Dibenzothiophenes (C_0-C_4)	1,330	18,713
Fluorenes (C_0-C_4)	348	12,800-30,700
Chrysenes (C_0-C_4)	168	21,600-48,700
Metals (ppm)		
Nickel	55	37
Vanadium	310	32
Zinc	19	45

SOURCE: Bitumenes Orinoco, S.A., b; National Oceanic and Atmospheric Administration ADIOS Model Database.

stabilizer, MEA. The purpose of these components is to maintain the stability of the bitumen droplets in the emulsion by preventing particle-particle agglomeration and coalescence.

The AE that make up the surfactant are composed of a long-chain fatty (alkyl) alcohol (hydrophobic) and an ethylene oxide (EO) chain (hydrophilic), connected by an ether linkage. The nomenclature of AE is determined by the average number of carbons in the alkyl chain of the alcohol and the number of EO groups in the hydrophilic moiety (e.g., $C_{10-12}EO_8$ represents an alcohol with 10 to 12 carbons attached to polyethylene oxide with 8 EO units).

The alcohol ethoxylate used in Orimulsion-400 is known as GENAPOL X 159 and is complexly branched. It contains a mixture of highly branched C_{12} (22-30 percent) and C_{13} (70-78 percent) fatty alcohols with anywhere from 9 to 22 (EO_9 to EO_{22}) EO groups (Bjornestad et al., 1998; Bowadt et al., 1998). The

TABLE 2.2 PAH Concentrations in Orimulsion-400, Crude Oils, and Petroleum Products

Sample Type PAHs	Average Light Crude (average of 19 oils) (µg/g oil)	Prudhoe Bay crude (µg/g oil)	Average Heavy Crude (average of 6 oils) (µg/g oil)	Arab. Med. Crude (µg/g oil)	Diesel No. 2 (µg/g oil)	Average Bunker C (average of 4) (µg/g oil)	Orimulsion (µg/g oil)	Cold Lake Bitumen (µg/g oil)
Naphthalene								
C0-N	**45.6**	46	**21.0**	21.0	1404.6	**28.2**	9.3	47.0
C1-N	**540.4**	540	**148.9**	148.9	5174.0	**87.6**	47.2	128.0
C2-N	**1696.9**	1697	**427.0**	427.0	7031.6	**327.8**	165.7	490.0
C3-N	**2082.9**	2083	**602.4**	602.4	5591.4	**541.0**	243.0	839.0
C4-N	**990.6**	991	**313.2**	313.2	2963.6	**507.0**	309	705.0
Sum	5356	5356	1513	1513	22165	1492	774	2209
Phenanthrene								
C0-P	**41.4**	41	**80.4**	80.4	455.8	**68.4**	45.7	91.0
C1-P	**348.3**	348	**324.2**	324.2	657.5	**210.2**	133.4	287.0
C2-P	**462.3**	462	**405.5**	405.5	223.5	**406.3**	306.5	506.0
C3-P	**380.3**	380	**356.4**	356.4	33.0	**472.2**	425.4	519.0
C4-P	**262.2**	262	**225.8**	225.8	6.3	**202.3**	228.6	176.0
Sum	1494	1494	1392	1392	1376	1359	1140	1579
Dibenzothiophene								
C0-D	**185.8**	186	**37.4**	37.4	367.4	**34.9**	16.7	53.0
C1-D	**623.5**	624	**124.9**	124.9	414.9	**136.8**	67.6	206.0
C2-D	**1092.9**	1093	**212.3**	212.3	160.5	**337.5**	222.9	452.0
C3-D	**967.5**	968	**192.5**	192.5	30.6	**405.0**	364.0	446.0
Sum	2870	2870	567	567	973	914	671	1157
Fluorene								
C0-F	**43.1**	43	**18.6**	18.6	179.4	**21.1**	10.1	32.0
C1-F	**120.3**	120	**60.6**	60.6	404.5	**54.8**	38.6	71.0
C2-F	**240.6**	241	**98.0**	98.0	375.0	**135.8**	126.7	145.0
C3-F	**225.8**	226	**109.3**	109.3	221.3	**152.2**	134.4	170.0
Sum	630	630	286	286	1180	364	310	418

Chrysene							
C0-C	**17.1**	17	**32.0**	32.0	0.5	**21.4**	14.9	28.0
C1-C	**26.3**	26	**82.9**	82.9	0.0	**39.8**	29.5	50.0
C2-C	**37.7**	38	**139.8**	139.8	0.0	**62.0**	53.9	70.0
C3-C	**34.5**	35	**72.5**	72.5	0.0	**44.6**	46.3	43.0
Sum	116	116	327	327	1	168	145	191
TOTAL PAH	10466	10466	4086	4086	25696	4297	3040	5554

Other PAHs							
Biphenyl	**26.9**	26.90	**2.7**	2.7	363.4	**11.2**	5.4
Acenaphthalene	**10.2**	10.22	**0.7**	0.7	31.7	**2.7**	1.6
Acenaphthene	**7.5**	7.50	**1.0**	1.0	24.5	**3.2**	11.2
Anthracene	**3.1**	3.06	**0.8**	0.8	59.5	**6.5**	2.7
Fluoranthene	**3.2**	3.19	**0.7**	0.7	0.1	**3.4**	3.2
Pyrene	**4.3**	4.26	**4.8**	4.8	0.3	**1.8**	6.5
Benz(a)anthracene	**0.9**	0.86	**5.2**	5.2	0.0	**0.3**	3.4
Benzo(b)fluoranthene	**1.3**	1.26	**3.2**	3.2	0.0	**2.2**	1.8
Benzo(k)fluoranthene	**0.4**	0.43	**0.7**	0.7		**2.1**	0.3
Benzo(e)pyrene	**3.1**	3.13	**12.2**	12.2	0.0	**6.7**	2.2
Benzo(a)pyrene	**0.5**	0.46	**2.4**	2.4	0.1	**0.3**	2.1
Perylene	**0.1**	0.09	**0.8**	0.8	0.0	**0.2**	6.7
Indeno(1,2,3cd)pyrene	**0.1**	0.06	**0.2**	0.2	0.1	**1.9**	0.3
Dibenz(a,h)anthracene	**0.2**	0.21	**1.4**	1.4	0.0	**49.5**	0.2
Benzo(ghi)perylene	**0.6**	0.58	**7.0**	7.0	0.0	**48.6**	1.9
TOTAL	62	62	44	43.8	480	141	49

Source: Environment Canada, 2001.

selected group of C_{12} and C_{13} branched AE was used in Orimulsion-400 rather than nonylphenol ethoxylates used in Orimulsion-100, because of the latter's potential for endocrine disruption and the fact that the degradation metabolites were more toxic and persistent than the parent compounds (Harwell and Johnson, 2000).

Because of their widespread use in household products, much of the research on microbial degradation of AE has been based on their removal efficiencies in wastewater treatment facilities. For various types of treatments, 86-99 percent of the AE in wastewater influents are degraded to some intermediate form (Talmage, 1994). It has long been held that the major, if not only, route for linear AE biodegradation is that of cleavage at the ether bridge between the alkyl chain and the EO moiety (Swisher, 1987). After that, biodegradations of fatty alcohols and polyethylene glycols (PEG) were believed to proceed independently and more slowly.

Branching of the alkyl chain in the vicinity of the central ether bridge appears to inhibit central ether cleavage (Di Corcia et al., 1998). In addition, the same study (Di Corcia et al., 1998) reported that bacterial attack on the ethoxy chain produced metabolites with the EO either shortened or, to a lesser extent, oxidized to a terminal carboxylic acid group. They also reported end-of-chain oxidation of both alkyl side chains in branched AE to form very polar di- and tri-carboxylic acids. Marcomini et al. (2000a,b) reported fast biomediated ether-linkage cleavage of linear and short-chain (methyl or ethyl) C_2-monobranched AE with slower biodegradation of the released PEG by both hydrolytic shortening and oxidative hydrolysis to form shorter PEG oligomers and carboxylated PEG. Taken together, these studies suggest that the AE mixtures used in Orimulsion-400 formulations are capable of slow aerobic biodegradation. In fact, one study on the rate of biodegradation of GENAPOL X 159 stated that "the toxicity of the AE was still 100 percent even after 56 days of biodegradation." This is in accordance with the results of Bjornestad et al. (1998) who found biodegradation of only 35 percent of the AE. This rate of biodegradation is at variance however with oxygen consumption studies on the specific AE mixture used for Orimulsion-400 that have shown a biodegradability (compared to complete chemical oxygen demand with potassium dichromate) of 79 percent in seawater over a 28-day period (VKI, 1997a).

MEA, the substance used to stabilize Orimulsion, is widely used in healthcare products and the surfactant industry. It is not considered mutagenic or carcinogenic (Johnson, 1998) and is utilized and metabolized by plant, animal, and microbial cells. As such, it is quickly transformed in the environment with a half-life of days to weeks (Johnson, 1998). Oxygen consumption studies on MEA have shown a biodegradability (compared to complete chemical oxygen demand with potassium dichromate) of 74 percent in seawater over a 28-day period (VKI, 1997b). Because of its high water solubility, MEA has a low potential for bioaccumulation. This is discussed in greater detail in Chapter 3.

TABLE 2.3 Selected Dissolved Aromatic Hydrocarbon Concentrations (Mean and Standard Deviation, µg/L) After Five-Day, Zero-Headspace Equilibrium Exposure Studies

Fuel	Benzene	Toluene	Xylene	Naphthalene
No. 6 fuel oil	73.0 ± 15.8	197.3 ± 29.2	126.5 ± 17.2	142.5 ± 23.6
Orimulsion	5.9 ± 4.2	57.8 ± 5.1	3.8 ± 1.6	4.9 ± 2.5

SOURCE: Brown et al., 1995.

A detailed chemical characterization of the water-soluble components contained in the 30 percent water added to make Orimulsion and the water-soluble fractions (WSF) generated by dispersions of Orimulsion into fresh and salt water was completed by Potter et al. (1997). In addition, Brown et al. (1995) examined the dissolution behavior of Orimulsion and No. 6 fuel oil spilled in water of varying salinities. These studies concluded that the majority of the surfactants are contained in the aqueous phase and that they would be diluted by the receiving water during an Orimulsion spill. BTEX constituents were not observed in 1:9 (volume:volume) dispersions of Orimulsion in water at a detection limit of 0.2 µg/L (Potter et al., 1997). Table 2.3 presents benzene, toluene, xylene and naphthalene concentrations from zero-headspace, five-day equilibrium exposure studies comparing Orimulsion and No. 6 fuel oil (Brown et al., 1995). In general, levels of volatile and water-soluble components are higher by up to one order of magnitude for the No. 6 fuel oil tested in their studies compared to Orimulsion.

To date, no detailed chemical analyses of the dissolved-phase PAH concentrations in the 30 percent aqueous phase of Orimulsion have been reported. Using equilibrium partition theory and the initial concentrations of PAH in neat Orimulsion, Stout (1999) and French (2000) calculated the initial concentrations of PAH expected in the aqueous phase of Orimulsion; these values are presented in Table 2.4. The highest concentrations of any PAH are for the naphthalenes, and they are only around 2 µg/L—in general agreement with the values measured by Brown et al. (1995), with initial concentrations for other PAH rapidly decreasing to values in the range of 0.1-0.8 µg/L. The sum of the PAH, or the total PAH (TPAH) with log K_{ow} values <5.6 is only 14.9 µg/L. Brown et al. (1995) examined the time-series kinetics of dissolution behavior from Orimulsion and No. 6 fuel oil, and concluded that little or no additional environmentally significant PAH dissolution occurs upon release of Orimulsion to the environment. In their dilution studies, however, they did not dilute the Orimulsion fuel enough to allow for continued dissolution of PAH from the bitumen phase. Instead, they concluded that the dissolved-phase components measured in their experiments were simply from the dilution of the already near-equilibrium concentrations of PAH in the initial aqueous phase. A similar conclusion was reached by Stout (1999) who examined the dissolution behavior of the high-molecular-weight PAH frac-

TABLE 2.4 Dissolved PAH Concentrations Estimated in the Aqueous Phase of Fresh Orimulsion

PAH	Molecular Weight (g/mol)	Log K_{ow}	Concentration in Bitumen (mg/kg)	Dissolved Concentration (μg/L)
Naphthalene	128	3.37	15.4	2.363
C_1 naphthalenes	142	3.87	43.03	2.263
C_2 naphthalenes	156	4.37	136.7	2.464
C_3 naphthalenes	170	5	189.37	0.886
C_4 naphthalenes	185	5.55	267.97	0.386
Biphenyls	154	3.9	5	0.247
Acenaphthylene	152	4.07	0	0
Acenaphthene	154	3.92	10.66	0.504
Dibenzofuran	168	4.31	5.42	0.111
Fluorene	166	4.18	13.52	0.366
C_1 fluorenes	181	4.97	57.39	0.286
C_2 fluorenes	196	5.2	184.39	0.562
C_3 fluorenes	211	5.5	272.13	0.436
Dibenzothiophene	184	4.49	28.19	0.393
C_1 dibenzothiophene	199	4.86	133.96	0.846
C_2 dibenzothiophene	214	5.5	345.13	0.553
C_3 dibenzothiophene	228	5.73	692.83	0.679
Phenanthrene	178	4.57	67.78	0.796
Anthracene	178	4.54	0	0
C_1 phenanthrenes-anthracenes	192	5.14	143.84	0.499
C_2 phenanthrenes-anthracenes	207	5.25	366.41	1.003
C_3 phenanthrenes-anthracenes	222	6	459.43	0.252
C_4 phenanthrenes-anthracenes	237	6.51	241.73	0.0446
Fluoranthene	202.3	5.22	0	0
Pyrene	202.3	5.18	0	0
Sum (K_{ow}<5.6)			2,286	14.964

SOURCE: Stout, 1999.
NOTE: Concentrations are calculated from the PAH composition of whole Orimulsion data from Stout (1999) and equilibrium partitioning (data provided by D. French McCay, A.S.A. Narragansett, RI) to estimate dissolved PAH composition (PAH concentrations listed as zero if log K_{ow} >5.6).

tion of Orimulsion-400 after 24 hours' exposure in benchtop 4-liter stirred-beaker experiments. Figure 2.5 presents a histogram plot showing the relative abundance of high-molecular-weight PAH in fresh Orimulsion and the dissolved constituents in the water generated by gently stirring a 1,700-mg/L dispersion of Orimulsion in fresh water for 24 hours. The PAH dissolved in the water are enriched in the more soluble low-molecular-weight two-ring PAH (naphthalene and its alkyl-substituted homologues), which were the predominant dissolved-phase components in the near-equilibrium 30 percent aqueous phase in the original Orimulsion fuel mixture (e.g., see Table 2.4). As discussed further in the section titled "Water Column Processes - Dissolution," the high-molecular-weight

PAH (e.g., phenanthrenes, dibenzothiophenes, and chrysenes) in the dispersed bitumen phase (Figure 2.5) can continue to dissolve to an appreciable extent given the extremely high surface-area-to-volume ratio of the dispersed bitumen droplets. However, this dissolution process would be masked at the dilution volume used in the experiment. In addition, the coalescence of bitumen droplets into larger agglomerates would tend to limit such behavior in the experimental apparatus used, while it would not be a factor in an open-ocean environment.

WATER COLUMN PROCESSES

To date, there have been only a limited number of smaller experimental spills and no accidental spills of Orimulsion. As a result, much of our knowledge regarding the behavior and fate of spilled Orimulsion is based on laboratory-scale studies and models. In preparing a "real-world" Orimulsion budget or mass balance based on predictions from benchtop or flume experiments, it is critical to caveat the results because of the difficulty in simulating and scaling "real-world" conditions. Factors that cannot be simulated in laboratory-scale experiments include water turbulence and density profiles, unconfined water volumes, continual dilution of bitumen, and the absence of container walls (Stout, 1999). Moreover, because Orimulsion is not a homogeneous fluid, there may be special peculiarities in its behavior and the breakdown of the emulsion that are specific to the nature of the spill event.

Dispersion

Orimulsion is expected to disperse (spread apart) rapidly when spilled into an open-water environment. In fact, for spill response purposes, fresh Orimulsion can almost be considered predispersed oil. The near-equilibrium dissolved-phase PAH in the aqueous phase of the product would behave as a neutrally buoyant, or near neutrally buoyant, contaminant. Dilution of this dissolved-phase PAH solution would depend on the mixing characteristics of the surrounding water. As an example, two small spills of Orimulsion were intentionally released into the Caribbean Sea in July 1996 (French et al., 1997). Using the reported diffusion coefficients for this experiment, neglecting boundary conditions and buoyancy forces, and assuming Fickian diffusion, the maximum dissolved-phase PAH concentration of the experimental spills would fall to less than 1 percent of the initial concentration in less than two minutes. However, eddy diffusion coefficients are functions of the flow field and vary in both time and space. For this reason measured concentrations can differ by orders of magnitude (Okubo, 1971). Actual dilution will therefore depend on the specific nature of the spill and the concurrent environmental conditions. Also, additional dissolution of PAH from the dispersed bitumen droplets will further complicate the picture as described in the following section. In general, more open and energetic conditions will result in more rapid mixing and lower concentrations.

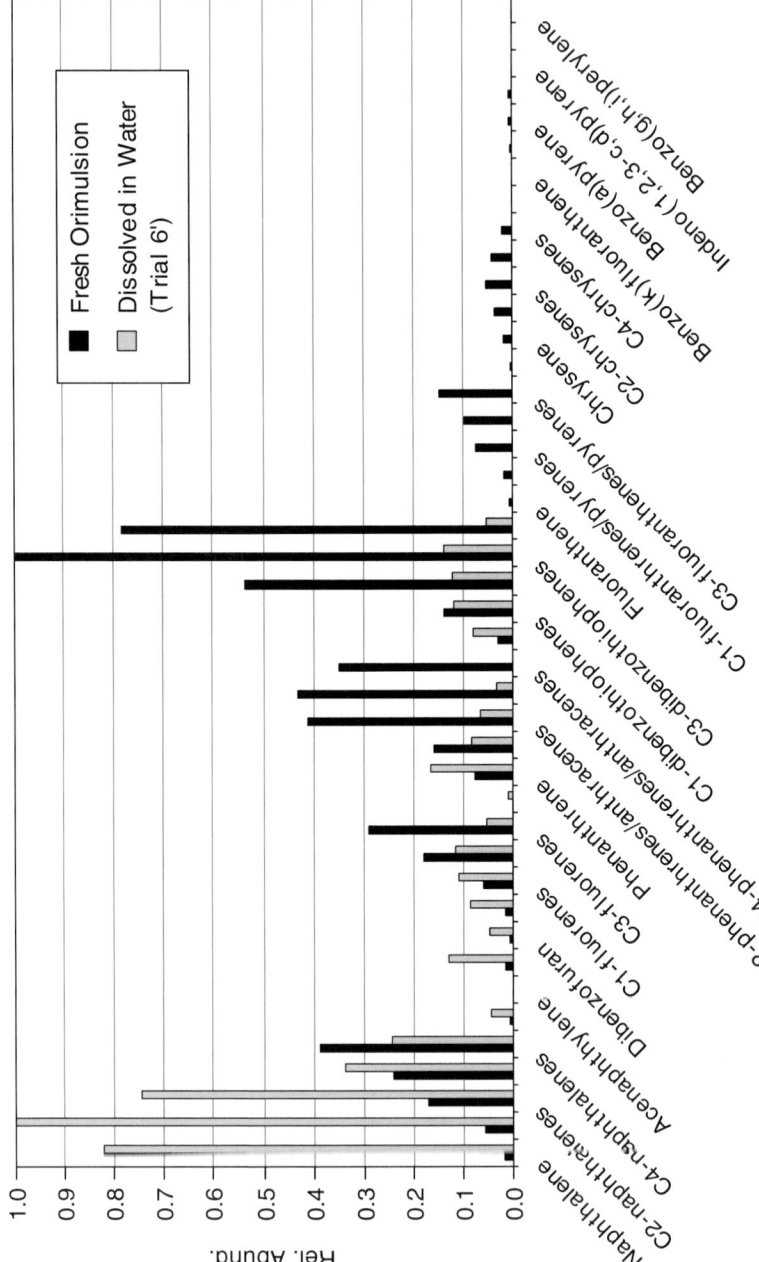

FIGURE 2.5 Histograms comparing the distribution of PAH in fresh Orimulsion and a representative distribution of PAH dissolved in water after 24 hours. NOTE: Because of space limitations, compound names are given only for every other PAH. SOURCE: Stout, 1999.

In these open-water experimental spills, the bitumen concentration quickly fell below the critical concentrations reported by Febres et al. (1995) and Brown et al. (1995), where there is a transition from an emulsion to a cloud of suspended bitumen droplets. In such a situation, the behavior of these droplets depends upon the initial concentration, the mixing energy of the system, and the salinity and temperature of the water. The initial size bitumen droplets in an Orimulsion spill are nearly all below the size limit that spill responders use to define dispersed oil (Lunel, 1993). According to Li and Garrett (1998), fluids with a high viscosity such as bitumen resist being broken by normal turbulence, so it is unlikely that the droplets would be further reduced in size. Small droplets will tend to remain suspended in the water column pending coalescence and the formation of larger agglomerates. Coalescence is discussed further in "Coalescence and Bitumen-SPM Interaction" later in this chapter. If there were no coalescence of the bitumen or adherence to suspended sediments, the droplets would be expected to behave in a similar manner as dust particles in the air. While dust particles are heavier than air, atmospheric turbulence keeps them suspended. Similarly, even minimal water turbulence would keep the bitumen droplets from either sinking (in fresh water) or floating (in seawater) as would be expected, based solely on density considerations. Coalescence, which would increase droplet size, could change this result because the vertical velocity of the bitumen droplet due to buoyancy increases with the square of the droplet diameter (small particles) or linearly with the droplet diameter (large particles).

In an area of steady currents, vertical shear diffusion would cause the bitumen plume to become elongated and progressively more dilute in the direction of the current. Shorelines or other boundaries could restrict the dispersion of the subsurface droplets but, to the extent that the relatively weak buoyancy forces can be neglected, would not cause any increased concentration of the bitumen. The exception would be any bitumen that has resurfaced or sunk due to coalescence.

Dissolution

An Orimulsion spill into a water body results in input of dispersed bitumen droplets and the dissolved constituents in the aqueous phase (BTEX, dissolved PAH at near-equilibrium levels, and surfactants). The initial and final concentrations of dissolved constituents will be a function of the volume of Orimulsion spilled, the volume of the receiving water body, and the turbulence.

Laboratory studies by Brown et al. (1995) have confirmed that the transfer of initially dissolved components from Orimulsion into water is essentially instantaneous, usually occurring within 30 minutes. This process is reflected in Figure 2.6, which shows the dissolution behavior observed when Orimulsion was introduced into seawater at varying salinities in controlled equilibrium batch studies. These curves document the behavior expected from simple dilution of the near-equilibrium dissolved PAH (see also Figure 2.5) and the surfactants in the origi-

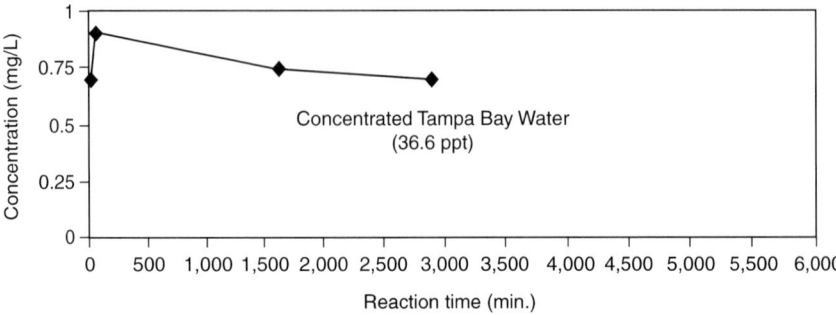
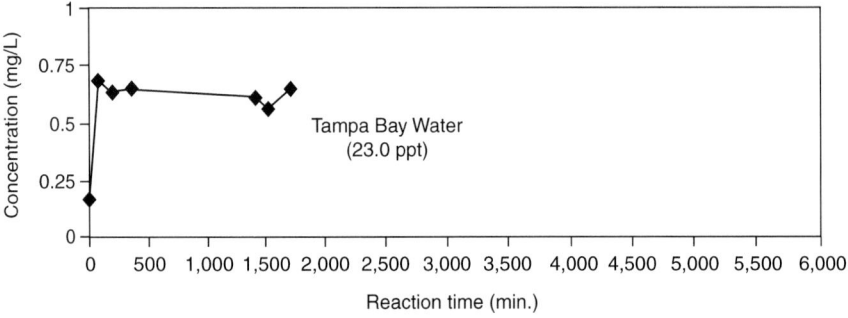
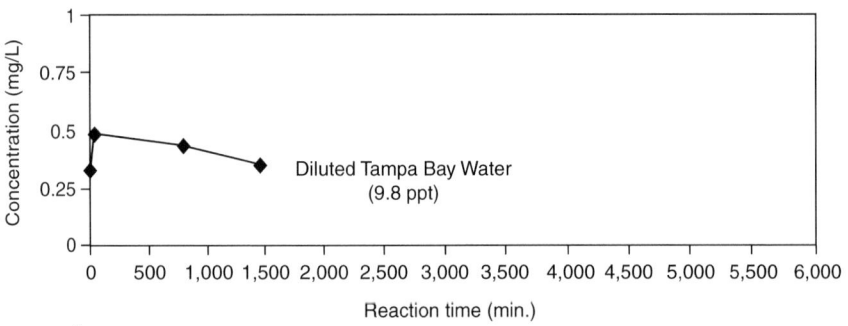

FIGURE 2.6 Mean values of total dissolved aromatic hydrocarbons from Orimulsion as a function of salinity and batch-mode equilibrium exposure time. SOURCE: Brown et al., 1995. Used with permission of the University of Miami.

nal Orimulsion aqueous phase. Interestingly, the laboratory data suggest that total dissolved aromatic hydrocarbon concentrations are actually higher at high salinity (36 ppt) compared to brackish water salinities (9.8 ppt). This behavior seems counterintuitive at first. However, Brown et al. (1995) believed that the increased concentrations in the dissolved phase at higher salinities might be caused by the preferential formation of micelles by free surfactants under high ionic strengths, with an apparent solubility increment resulting from the incorporation of petroleum hydrocarbons into the surfactant-based micelles. Although this hypothesis seems reasonable, additional studies are needed to fully evaluate and confirm it.

More importantly, however, a number of researchers interpreted these data to conclude that continued PAH dissolution from the bitumen droplets would not be environmentally significant. What these studies failed to consider was the limited dilution volume of the receiving water, the influence of bitumen droplet coalescence on droplet size distribution, and the importance of equilibrium partitioning theory. Equilibrium partitioning theory predicts that additional PAH should dissolve from the bitumen droplets (especially given their high surface-area-to-volume ratios) when the water phase is sufficiently diluted. These topics are considered in more detail below; but first, by way of comparison, Figure 2.7 shows the dissolution behavior of PAH components when No. 6 fuel oil was introduced into the controlled batch-mode equilibrium studies completed by Brown et al. (1995). Unlike the Orimulsion tests, No. 6 fuel oil continues to release dissolved-phase PAH into the water over time even under the rather limited dilution conditions of the laboratory tests. The observed increased concentration of the dissolved fraction with decreasing salinity is expected given the solubility behavior for hydrophobic organic compounds in seawater.

Once diluted into the receiving body of water, dissolved-phase PAH, AE, and monoethanolamine would be subject to loss by bacterial degradation at rates controlled by bacterial population densities, nutrient levels, temperature, and other factors. Adsorption of dissolved-phase PAH by SPM is not expected to be significant compared to direct bitumen droplet-SPM interactions. Stout (1999) concluded, like Brown et al. (1995), that the PAH components associated with the bitumen droplets would be essentially inert, with very little additional abiotic (evaporation-dissolution) or bacterial weathering expected. In more recent theoretical and modeling evaluations of Orimulsion behavior, French (2000) concluded that significant PAH dissolution weathering from the bitumen droplets can occur over time, particularly given the high surface-area-to-volume ratios of the 1- to 80-μm-diameter dispersed bitumen droplets (Stout, 1999).

French, using equilibrium partitioning theory, and the initial PAH concentrations in neat Orimulsion from two different sources determined that the effective initial concentration of dissolved PAH in the neat fuel would be in the range of 15 μg/L (calculated as shown in Table 2.4 based on the 2,286 ppm of PAH in neat Orimulsion reported by Stout (1999), and 30 ppb (calculated in a similar manner but using the 3,050 ppm of PAH in neat Orimulsion as reported in Table 2.2 by Environment Canada (2001). This represents the maximum theoretical concentration that could be seen by exposed organisms. Concentrations (μg/L) of dissolved PAH and monoaromatic hydrocarbons (MAH) from Orimulsion diluted

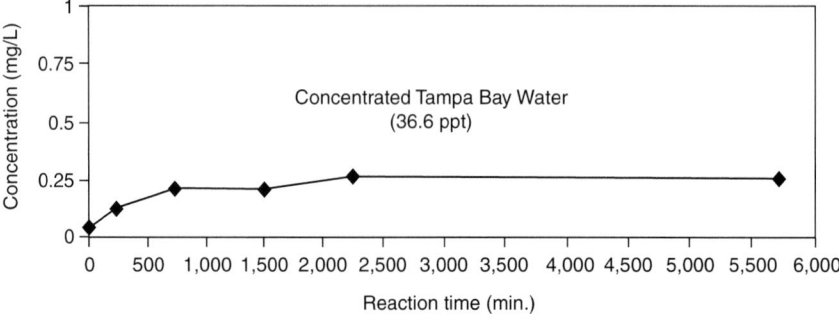

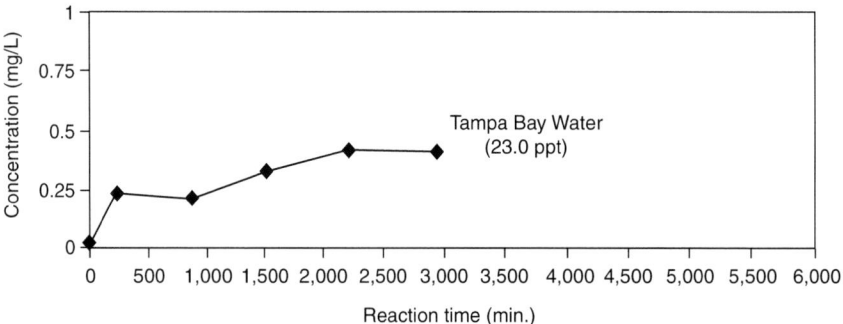

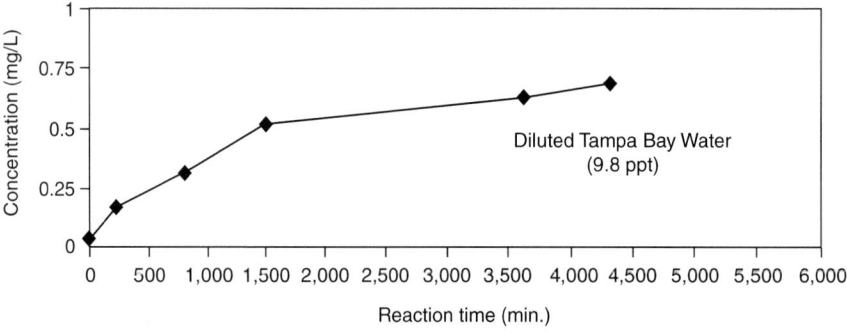

FIGURE 2.7 Mean values of total dissolved aromatic hydrocarbons from No. 6 fuel oil as a function of salinity and batch-mode equilibrium exposure time. SOURCE: Brown et al., 1995. Used with permission of the University of Miami.

by the indicated dilution factors were calculated using the equations of French and are shown in Table 2.5. As shown by the data in the table, PAH are predicted to continue to dissolve from the dispersed bitumen droplets as they are diluted into the receiving water body, but the dissolved PAH will never exceed the initial

TABLE 2.5 Concentration of Dissolved Aromatics (µg/L) from Orimulsion Diluted by the Indicated Factors, as Computed from an Equilibrium Partitioning Model

Dilution	MAH + PAH	MAH	PAH
1	51.06	32.05	19.02
10	50.64	31.64	19.01
70	48.25	29.29	18.95
1,00	47.24	28.31	18.93
1,000	34.52	16.30	18.22
10,000	18.43	3.93	14.49
100,000	7.84	0.47	7.37
1,000,000	2.01	0.05	1.96

SOURCE: Data provided by D. French McCay, A.S.A., Narragansett, RI.

concentrations in the aqueous phase of the emulsion. Increased dissolution with exposure to noncontaminated water can keep up with dilution initially (up to approximately a thousandfold dilution); however, after that, calculated dissolved-phase concentrations decline due to dilution. The concentration at 1:70 dilution agrees with the lowest salinity result by Brown et al. (1995), as one would expect because the partition coefficients used here are for measurements in fresh water.

The more dilute the fuel, the higher is the percentage dissolved (Table 2.6), such that the dissolved concentrations do not represent simple dilutions of the originally dissolved fraction in the neat fuel. The Brown et al. (1995) results indicate that dissolution is very rapid and is complete in less than one hour (at least at 1:70 dilution).

TABLE 2.6 Percentage of MAH and PAH Dissolved from Orimulsion Diluted by the Indicated Factors, as Computed from an Equilibrium Partitioning Model

Dilution	MAH + PAH	MAH	PAH
1	0.00	0.07	0.0010
10	0.02	0.66	0.010
70	0.11	4.24	0.07
100	0.16	5.86	0.10
1,000	1.16	33.75	0.91
10,000	6.22	81.39	7.17
100,000	26.46	97.70	34.62
1,000,000	67.76	99.76	80.00

SOURCE: Data provided by D. French McCay, A.S.A., Narragansett, RI.

Coalescence and Bitumen-Suspended Particulate Matter Interaction

Because there have been no large accidental releases of emulsified bitumen, one can only speculate about the behavior of such a spill. The open-ocean turbulent energy spectrum and lack of boundaries are not easily reproduced in the laboratory. There could be other phenomena that appear only under full-scale conditions and that do not occur in laboratory studies or small field experiments. This uncertainty must be considered in describing expected bitumen coalescence and other Orimulsion spill issues.

While the surfactant remains effective, there will be little coalescence of the bitumen droplets, in either fresh or salt water. However, salinity as low as 6 ppt can collapse and deactivate the surfactant. Specifically, dissolved salt dehydrates the EO sheath and compresses the electrostatic double layer around the bitumen droplets (Crosbie and Lewis, 1998a).

The proportion of bitumen that disperses or coalesces will depend on the spill volume and conditions that prevail at the time of an Orimulsion spill. Closed-system flume studies (Ostazeski et al., 1998a, 1998b), using seawater with breaking waves, produced a rapid increase in the mean bitumen droplet size. These larger droplets scavenged other bitumen droplets as they rose to the water surface, forming floating tar patties. Minimal coalescence occurred in similar tests using fresh water, presumably due to the higher effectiveness of the surfactant in fresh water. In laboratory studies in both brackish and salt water, the majority of the bitumen ended up floating on or near the surface.

In salt water, turbulence contributes to two competing phenomena that affect the dispersed bitumen. On the one hand, the turbulence increases dispersion of the bitumen droplets. On the other hand, it also contributes to particle-particle collisions and coalescence leading to the formation of floating tar mats on the water surface. Stout (1999) predicted that the propensity for droplet coalescence is probably greater under lower-energy conditions. Anecdotal reports from observers at small, open-water experimental spills indicate that as much as one-third of the bitumen will coalesce and surface. Higher-energy situations will lead to small tarballs, whereas lower-energy settings are more likely to produce larger tar patties or ropes. Spill volume and release rate will also be key factors, because larger instantaneous spills will have higher droplet concentrations and rates of interaction. This is an important area for future studies because the percentage of bitumen that surfaces or remains in the water column affects the method of cleanup and expected environmental impacts of the spill. For purposes of this report, an assumption has been made that coalescence and surfacing of the bitumen in salt water and brackish water will be significant.

Interactions with Suspended Particulate Material

Numerous batch-mode and flume studies have been undertaken to evaluate the interaction of dispersed bitumen droplets and suspended sedimentary material

(Bitumenes Orinoco, S,A., a; Brown et al., 1995; Johnson et al., 1998; Stout, 1999). Brown et al. (1995) examined the interaction of Orimulsion bitumen with fine and coarse sediment in saline water and reported that adhesion equilibrium reached up to 1 and 2.5 mg of bitumen per gram of coarse sediment at low and high sediment loads, respectively. Adhesion approached maximum values of 2,000 and 6,000 mg of Orimulsion bitumen per gram of fine sediment in batch equilibrium experiments. Orimulsion interactions with SPM are physical adhesion processes that seem to be favored by high salinity, sediment surface area, and/or high organic carbon content (Brown et al., 1995). In comparing the behavior of Fuel Oil No. 6 and Orimulsion, Brown et al. (1995) reported that the adhesion of No. 6 fuel oil to Tampa Bay sediments was negligible compared to Orimulsion. For 15-minute contact experiments, the fine sediment fraction showed approximately 1,400-mg/g loadings for Orimulsion and only 6 mg/g for No. 6 fuel oil. The coarse fraction loadings were 3.3 mg/g and 0.03 mg/g for Orimulsion and No. 6 fuel oil, respectively. Under brackish or full-strength seawater salinities, dispersed bitumen-SPM interactions can lead to formation of bitumen-SPM agglomerates that can be transported to the bottom (Brown et al., 1995; Stout, 1999). In flume studies conducted under extremely high-energy and full-strength seawater salinity conditions, up to 11 percent of the total bitumen was observed to be transferred eventually to the bottom in water containing high (45 mg/L) suspended loads of kaolinite (Stout, 1999). Similar experiments conducted in fresh water showed no transport of bitumen-kaolinite agglomerates to the bottom, with the majority of the dispersed bitumen remaining suspended in the water column, presumably because of the effectiveness of the surfactant in fresh water.

Long-Term Fate and Microbial Degradation of Sedimented Bitumen Droplets

It appears that Orimulsion is capable of being biodegraded although the rates are believed to be extremely slow. Brown et al. (1995) observed overgrowth of microbial life in Orimulsion weathering reactors containing Tampa Bay seawater, and Lapham et al. (1999) confirmed the slow aerobic biodegradation of Orimulsion bitumen, with 1-3 percent degraded over a 21- to 60-day period. Compared to degradation rates for crude oil, where 10 percent of the alkanes and 50 percent of the PAH were degraded over similar time scales, it is clear that the rates for bitumen degradation are extremely low. However, the authors concluded that it was remarkable that any microbial respiration of the highly degraded bitumen droplets occurred at all. They also observed that bitumen degradation was greater in the presence of the AE surfactant, which was believed either to act as a co-metabolite providing a readily available carbon and energy source to the bacteria or merely to increase bitumen reactivity by promoting its dissolution in water. In a subsequent study, Proctor et al. (2001) found that the addition of seagrass and pinfish to sediment microcosms stimulated the in situ degradation of bitumen by as much as two- to ninefold. This suggests that bioremediation augmented by the addition of natural marine carbon substrates may be a viable option for responding to spilled Orimulsion in the marine environment. Their studies also demonstrated that the bacteria were not growing on Orimulsion but were simply respiring it as a co-metabolite.

SHORELINE PROCESSES

Orimulsion Spills on Land and Shorelines

Fresh Orimulsion spilled directly onshore initially behaves similarly to heavy oils of like viscosity. It will tend to penetrate deeper into beaches that are wet and have gravel substrates. However, the mobility of Orimulsion decreases rapidly as it weathers and comes in contact with dry substrates, which could reduce the contamination threat to groundwater (AEA Technology, 1996). When spilled onto sand, the fresh emulsion is filtered, breaking the emulsion. The particles can penetrate approximately 4-5 cm before they occlude the available pore space, resulting in a maximum loading of approximately 55-60 percent bitumen in the surface layer on the contaminated sand (Harper and Kory, 1997). The water phase is available for percolation into the substrate, and the potential for groundwater contamination will be a function of local geology and spill size.

When floating bitumen strands on shorelines and dries, it becomes stickier and will not resuspend with the tide. Weathered bitumen, which is significantly stickier than fresh Orimulsion, will not penetrate sand as readily and is expected to initially penetrate only the coarsest beach sediments (cobble-boulder). With surface warming of the bitumen-coated substrates by solar radiation, however, the weathered bitumen may become fluid enough to percolate deeper into sediments (Harper and Kory, 1997).

Stranding of Weathered Bitumen

As long as bitumen that is dispersed in the water remains wet within the sediments, it will penetrate freely into pebble beaches but will remain on or near the surface of sand beaches (Harper and Kory, 1997). Unlike weathered bitumen patties, such dispersed bitumen is more mobile within the coarse-grained beach substrate than a typical heavy fuel oil. Dispersed bitumen can be flushed from sediments, but normal tidal flushing may not provide sufficient energy to remove it completely. Once the bitumen droplets have been exposed to air, they become much stickier and can form a tenacious coating on the surface of the beach sediments. This coating is difficult to remobilize, even if there is subsequent rewetting of the bitumen.

Bitumen that has already formed weathered patties or ropes on the water surface before reaching the shoreline is highly viscous and sticky, with many of the same characteristics as weathered fuel oils. It is unlikely to penetrate into sediments finer than pebbles (Harper and Kory, 1997). Sticky bitumen patties can mix with sand suspended by waves. Mixing with more than 1 percent suspended sand can cause the droplets to become heavier than typical seawater and sink.

USE OF MODELING AND SCENARIOS TO UNDERSTAND THE BEHAVIOR OF ORIMULSION SPILLS

A widely used oil spill model has been modified to simulate possible Orimulsion behavior (French and Mendelsohn, 1995). This model has been used to evaluate the effects of hypothetical Orimulsion spills in Tampa Bay and Delaware Bay and River (French McCay and Galagan, 2001) and as an analysis tool in a series of Orimulsion test spills in the Caribbean (French et al., 1997). Although the model incorporates standard spill model approximations and assumptions, certain parameters specific to Orimulsion were estimated from laboratory or small field experimental data. One such key parameter is the rate of bitumen coalescence, which was estimated by doing an empirical curve fit to bitumen concentration (French, 2000). Since this curve fit uses confined sample data, it is most likely too large. Although the algorithms of the model have been published (French and Mendelsohn, 1995; French et al., 1996), the committee was unable to verify that the numerical code in the model accurately represents these algorithms.

Other spill models have likewise been modified to simulate Orimulsion spill behavior using somewhat different approximations (VKI, 1999) and would presumably provide different predictions, although no direct comparison of the different models has been done. Barring an actual spill event, such model comparisons provide guidance on the sensitivity of selected environmental and computational parameters in model forecasting and may be an area of future research. Because each spill is unique, even an actual spill would only validate a model involving those particular circumstances. Nevertheless, modeling an emulsified bitumen fuel spill in different scenarios can be useful in sensitivity analysis and in determining where additional data are required. French and Mendelsohn (1995) used their model to assess the sensitivity of the model predictions to key factors such as bitumen coalescence. They also performed comparative studies of hypothetical Orimulsion and Fuel Oil No. 6 spills. They concluded that a large variety of scenarios are necessary to span the range of possible effects of emulsified bitumen spills, both in an absolute sense and in a relative sense for comparison with spills of other oil products. The committee selected six distinct spill scenarios to assess the expected behavior and fate of an Orimulsion spill.

The scenarios developed by the committee were (1) marine—open water, (2) marine—nearshore, (3) estuarine (brackish water), (4) nontidal river, (5) fresh water—quiescent, and (6) on land near water. The emphasis of these scenarios is on the behavior and fate of the bitumen component because of its unique properties and behavior once spilled into water. The dissolved constituents are also important, but the processes of mixing and dilution of water-soluble compounds is well understood.

Marine—Open Water

Figure 2.8 presents the schematic fate diagram for open-water marine spills of Orimulsion. Away from the immediate vicinity of an open-ocean spill, the surfactant will be diluted or degraded rapidly so that the emulsified fuel will become an independent cloud of bitumen droplets. The dissolved components in the water phase will mix quickly into the surrounding water. In open-water settings, concentrations should rapidly decrease due to mixing and turbulence. Some of the bitumen will coalesce and make its way to the surface as either tarballs or tar patties. The remaining bitumen will generally disperse due to water turbulence. The surface tarballs or patties will be transported by winds and surface currents, in a fashion similar to tarballs from any weathered fuel oil slick. Being only slightly buoyant, they will be subject to overwash, making their observation by spill response personnel difficult. Tarball fields can be spread over long distances and subsequently become reconstructed in convergence zones. Thus, a line of stranded tarballs from the spill could appear on beaches far away from the spill site. Upon exposure to the air, the surfaced bitumen will greatly increase its adhesion properties and could attach itself to any floating material it encounters. Mesocosm roof-top experiments have shown that prolonged exposure to sunlight can eventually lead to sinking of surface bitumen patties or

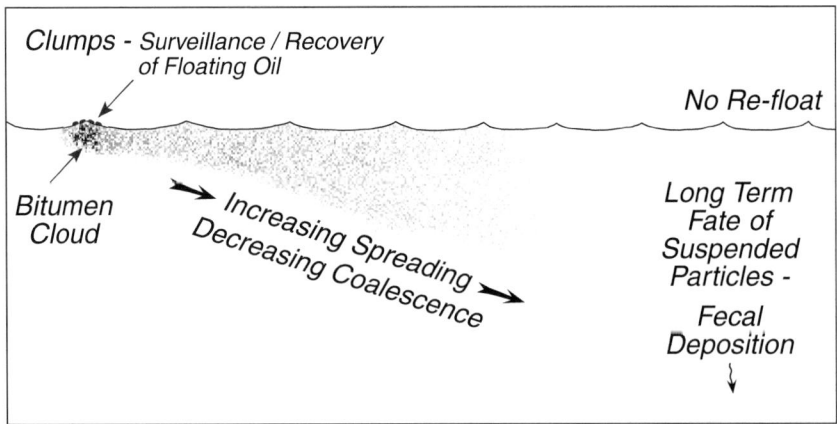

FIGURE 2.8 Schematic representation of the distribution and fate of Orimulsion-400 spilled into an idealized marine open water environment.

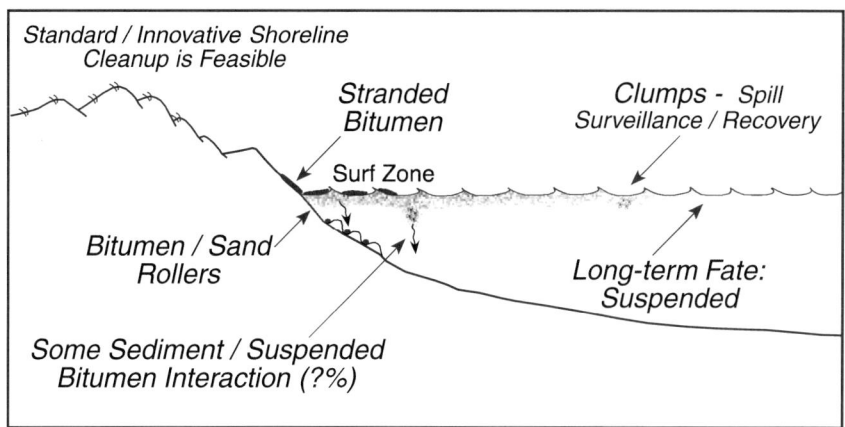

FIGURE 2.9 Schematic representation of the distribution and fate of Orimulsion-400 spilled into an idealized marine nearshore environment.

tarballs (Brown et al., 1995). Suspended bitumen and tar particles may also be subject to zooplankton grazing and, ultimately, sedimentation in fecal material.

Bitumen droplets are of a size range similar to single-celled algae (i.e., phytoplankton) and natural silt particles, which are fed on by filter feeders. However, most filter feeders studied to date have proven to be quite selective in their choice of food, and they can reject mineral and other particles. Copepods have been known to ingest droplets of Bunker C fuel oil following an oil spill (Conover, 1971), and up to 10 percent of the hydrocarbons in the water column were associated with the plankton and their feces. Laboratory studies with crustacean zooplankton, primarily copepods, demonstrated uptake of a variety of aromatic and paraffinic hydrocarbons from oil-contaminated food or water (Corner et al., 1976a, b; Lee, 1975); however, there are no data to show if similar behavior might be expected from ingested bitumen droplets.

Marine—Nearshore

Figure 2.9 presents the schematic fate diagram for nearshore coastal marine spills of Orimulsion, which are expected to behave differently than open-water spills. Because of the land and bottom boundary conditions, vertical diffusive mixing could be reduced and pockets of higher near-surface concentrations of bitumen and dissolved components could persist longer than would occur in open water. This would encourage a greater rate of coalescence and the formation of surface slicks. If the bitumen does resurface in quiescent water, it will tend to form a thin film of less than 0.1 mm (Sommerville, 1999) and can easily be resuspended if sufficient energy becomes available. Surface slicks of weathered bitumen can also become stranded at the high-water mark on the shoreline during a falling tide and onshore winds. Once deposited, they would be expected to weather as described in the previous section on shoreline interactions. If large patties or ropes of the weathered bitumen mix with sand suspended in the surf zone, they may form large "sand rollers," which are mixtures of tar and sand that can roll down and populate the nearshore bottom. Once deposited, they may be subject to burial in offshore bars. They could also become buried in the intertidal zone during depositional phases on beaches.

Suspended bitumen droplets would probably not strand on the beach but would be transported along the shoreline by the alongshore current. They could, however, be scavenged by suspended particulate matter as described earlier. In any segment of coastline that is of lower energy (or in calmer offshore waters), these bitumen/suspended sediment combinations could sink. Suspended or dispersed bitumen droplets that are carried offshore would be subject to the same long-term fate as bitumen droplets produced from an open-ocean Orimulsion spill.

Estuarine (Brackish Water)

Figure 2.10 presents the schematic fate diagram for an Orimulsion spill in estuarine or brackish water environments. Although the surfactant in Orimulsion will degrade in water with salinity greater than 5-7 ppt, the dispersed bitumen droplets will typically not become buoyant until the salinity of the surrounding water is 15 ppt or greater (Deis et al., 1997). Thus, it is possible for the bitumen droplets to begin to coalesce and then either float to the surface, sink to the bottom, or remain neutrally buoyant, depending on the encompassing water den-

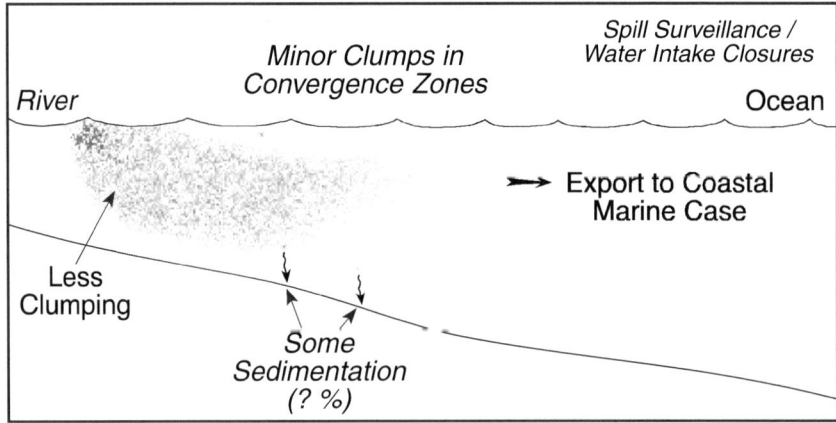

FIGURE 2.10 Schematic representation of the distribution and fate of Orimulsion-400 spilled into an idealized estuarine (brackish water) environment.

sity. If the water is stratified and quiescent, it is possible that increased concentrations of bitumen could form at the boundary between the less dense and more dense water. A likely scenario for an Orimulsion spill in a brackish, low-energy estuary of varying salinity would be that a small amount of the bitumen would coalesce and float to the surface, some would be scavenged by suspended particulates, and most would either sink very slowly to the bottom or remain in the water column, gradually being flushed by the normal water exchange in the estuary. Dissolved components (PAH, surfactants) are expected to decrease at rates determined by the degree of tidal mixing and flushing.

If the water is sufficiently brackish to cause some of the bitumen to float, this weathered floating bitumen could adhere to estuarine vegetation and impact shorelines. Bitumen/particle agglomerations could settle into the bottom sediment in quiescent areas where fine sediment accumulation occurs. If there is insufficient energy to cause coalescence, any buoyant bitumen may float to the surface but be subject to resuspension if disturbed. Suspended and dispersed bitumen droplets that are carried out of the estuary to coastal or offshore waters would be subject to the same long-term fate as bitumen derived from an open ocean Orimulsion spill.

Nontidal River

Figure 2.11 presents the schematic fate diagram for an Orimulsion spill in a nontidal energetic riverine environment. A river spill of Orimulsion would typically result in rapid mixing of the bitumen and dissolved components throughout the water column with subsequent rapid dilution. The surfactant would usually remain effective for a sufficient period of time to allow the bitumen droplets to

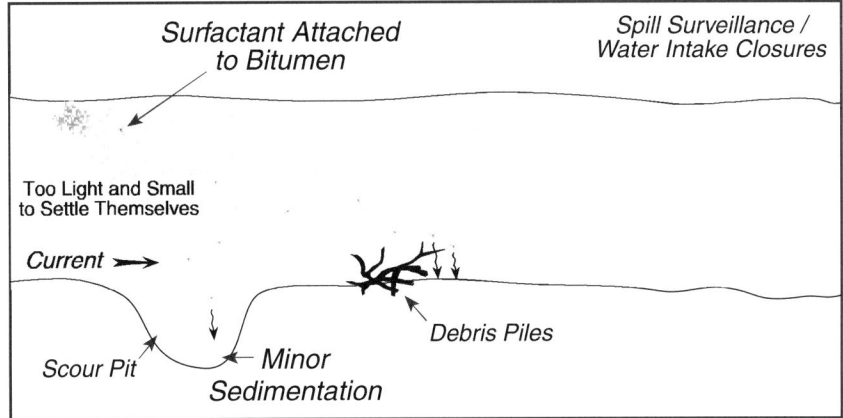

FIGURE 2.11 Schematic representation of the distribution and fate of Orimulsion-400 spilled into an idealized nontidal river environment.

become effectively dispersed to low enough particle densities to discourage coalescence. If there is a heavy sediment load in the river, there could be a minor amount of scavenging of the bitumen, although laboratory flume experiments have indicated that bitumen-SPM interactions are insignificant in fresh water conditions. There would most likely be little sinking of the pure bitumen outside of any previously formed scour pits or other naturally quiescent pools, where the droplets might temporarily collect but could easily be resuspended. Water intakes would have to be closed until the cloud of droplets passed. Flood conditions, a not uncommon situation in spill accidents, could strand bitumen-laden water on inundated floodplains. As the water drained off, the bitumen could remain on the top sediment layer. The long-term fate under normal conditions may be deposition in calm deltaic areas where other fine sedimentary material accumulates, although there are no data to support this hypothesis.

Fresh Water—Quiescent

Figure 2.12 presents the schematic fate diagram for an Orimulsion spill into a freshwater region with low or near-zero currents. A spill of Orimulsion into a quiescent freshwater pool would be similar to a spill in a river with less mixing and essentially no bitumen-SPM interaction. As in the river spill, the surfactant would probably remain effective long enough to allow the bitumen droplets to diffuse to a low enough concentration to inhibit coalescence, as long as the receiving water was of sufficient volume to allow breakdown of the emulsion. However, the low mixing energy would delay this process. The PAH fraction in the carrier aqueous phase would be subject to limited dilution if the volume

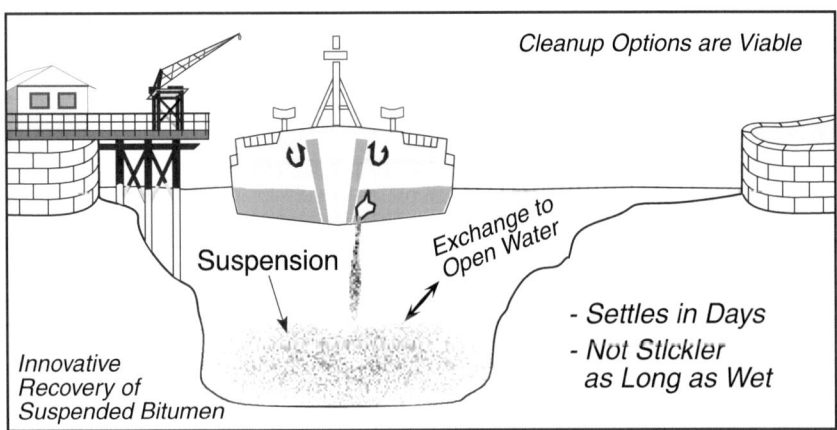

FIGURE 2.12 Schematic representation of the distribution and fate of Orimulsion-400 spilled into a quiescent fresh water environment.

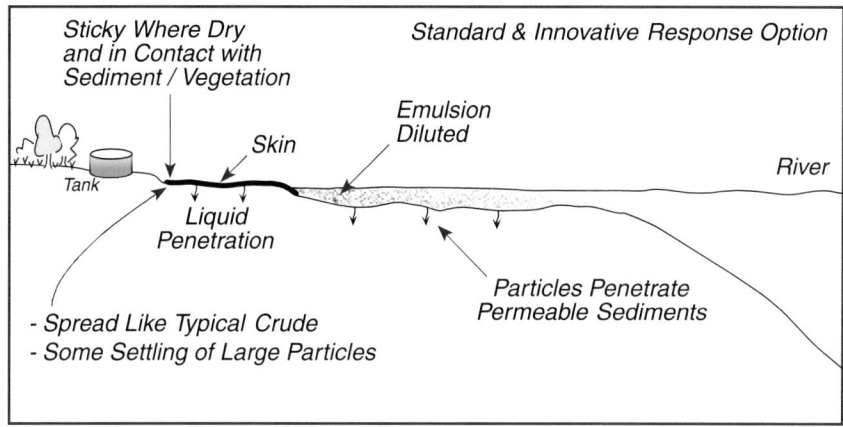

FIGURE 2.13 Schematic representation of the distribution and fate of Orimulsion-400 spilled into an idealized land-to-water (through wetlands) environment.

fraction of the Orimulsion were significant compared to the freshwater pool volume. Given a breakdown of the emulsion, some of the bitumen could settle temporarily to the bottom, but it could also be resuspended easily. If there were any turbulent energy, the bitumen droplet concentration would be nonzero throughout the water column, with the settling motion of the bitumen offset by random particle motion. For a spill on the surface of a pond, the steady-state vertical concentration of droplets could be an exponentially decaying function upward from the bottom of the pond (Hemond and Fechner, 1994).

On Land Near Water

Figure 2.13 presents the schematic fate diagram for an Orimulsion spill on land with subsequent migration through a wetland into a riverine environment. If Orimulsion is released on land, it would be expected to behave initially like a fuel oil with a similar viscosity. Penetration of fresh Orimulsion into soil and sandy sediments would be limited to the upper few centimeters, as described in the discussion of Orimulsion spills on land and shorelines. The suspended bitumen droplets are filtered out of the spilled material by the sand or soil, whereas the dissolved-phase surfactants and low-molecular-weight PAH constituents may percolate deeper into the soil and could eventually interact with groundwater. With prolonged exposure to air, the bitumen droplets remaining in the upper soil layers would dry, increasing their stickiness and viscosity, and any pooled Orimulsion would eventually turn into tar mats. If sufficient Orimulsion were spilled to eventually reach a freshwater body, some of the bitumen would be subject to dispersion in the receiving water body. However, under calm conditions, it would most likely behave like a highly viscous, high-density oil and settle toward the bottom in tar patties, tarballs, or tar mats, depending on the spill and environmental circumstances. The dissolved PAH fraction in the aqueous phase would be subject to limited dilution, and it could remain in the receiving waters at higher concentrations than in any of the other scenarios considered.

SUMMARY OF THE BEHAVIOR AND FATE OF SPILLED ORIMULSION

All of our understanding of the behavior of spilled Orimulsion is based on small-scale laboratory studies, flume tests, small (2-10 barrels) experimental spills in harbors and open water, and computer models developed from these data. Therefore, the ability to predict what happens in a real, large spill event remains limited, and responders should be prepared for a wide range of possibilities for response and cleanup. **To provide better prediction tools during spills, the different Orimulsion models should be compared for the same spill scenarios and the strengths and weaknesses of each model should be evaluated.**

Orimulsion behavior varies significantly when spilled into fresh water or into salt water due to the denaturing of the surfactant when the salinity of the receiving water is greater than 5-7 ppt. As a result, there is a greater tendency for dispersed bitumen droplets to coalesce and surface in brackish or salt water. The competing processes of coalescence (with possible surfacing or sinking) and dispersion into the water column dictate the behavior of bitumen droplets from an Orimulsion spill in marine or brackish water. **Therefore, further research must be done to quantify the processes of coalescence and dispersion of the bitumen droplets. In particular, the role of turbulent energy (magnitude and structure), salinity, spill volume, and spill rate should be evaluated.**

For spills into open water, Orimulsion would quickly behave as a cloud of dispersed bitumen droplets that are separated from the dissolved PAH and surfactants in the 30 percent water phase, which quickly mixes into the receiving water body. The bitumen droplets chemically resemble the residue that would be found in a heavy fuel oil spill at the end of short-term weathering processes. The concentrations of dispersed bitumen droplets and water-soluble constituents decrease significantly due to dispersion into the surrounding water column. Predictions of the dispersion are dependent on knowledge of the vertical and horizontal diffusion parameters that are key factors used in computer models. **To improve the accuracy of model predictions in support of local spill response plans, site-specific studies are needed to define diffusion coefficients (energy dissipation rates) for areas where Orimulsion shipment and/or loading and offloading operations are planned.**

The PAH concentration and chemical composition of bitumen are similar to the end product or heavily weathered residues from most crude oil spills. Significant quantities of the water-soluble PAH constituents have already leached from the bitumen droplets. Long-term leaching of PAH can occur at appreciable rates given the high surface-area-to-volume ratios of these 1- to 80-µm-diameter bitumen droplets. However, given the relatively low concentration of the high-molecular-weight PAH in the bitumen itself, this continued leaching is not expected to be environmentally significant. Limited data suggest that bitumen droplets may interact with SPM to a greater extent than No. 6 fuel oil (which does not break up into small droplets and remains suspended in the water column). Because the bitumen droplets are recalcitrant and remain in the environment for a long time, their interactions with SPM and retention in nearshore, estuarine, and riverine sediments may be important. **Additional dispersed bitumen-SPM studies using a wider variety of suspended particulate material types (including organic-rich substrates) at different salinities are recommended.**

The ultimate fate of spilled bitumen droplets is sedimentation, where continued biodegradation of the bitumen droplets is expected to be extremely slow. Based on the highly weathered nature of bitumen, the bioavailability of PAH is likely to be low. However, given the high surface-area-to-volume ratio of the very fine bitumen droplets, the potential for PAH uptake exists and thus should be evaluated.

For Orimulsion spills on land, the water phase can separate from the bitumen and infiltrate the underlying substrate. Because low- and intermediate-molecular-weight PAH and surfactants reside in the aqueous phase, they may reach groundwater. The potential for groundwater contamination is very site specific and outside the scope of this study.

Recent literature shows that the surfactant AE are subject to a variety of microbial breakdown pathways to highly oxygenated water-soluble intermediates. Identified breakdown products include fatty acids, alcohols, carboxylated polyethylene glycols, and (for branched AE) intermediate compounds with carboxylic acid groups on one or more alkyl groups (side chains) as well as the terminal end of the polyethylene oxide moiety. Branched AE such as those found in Orimulsion degrade slowly, and intermediate products can still be isolated (at low concentrations) from laboratory experiments after weeks to months. Therefore, AE and their intermediate degradation compounds can persist for weeks to months after a release. **Federal and state agencies should consider developing information on the ambient concentration of these compounds and their degradation products in the environment.**

3

Ecological Effects: Summary and Evaluation of Available Information

As pointed out earlier, most spill assessments are directed toward spills of floating oil. Thus, as with aspects of spill behavior and spill response discussed in the previous chapters, this chapter often compares and contrasts the potential ecological effects of spills of emulsified fuels with those of floating oils.

The ecological effects associated with spills of floating oils (including crude oil and distilled products) can generally be categorized as: (1) physical effects associated with smothering or coating on birds, mammals, or other organisms; (2) acute lethal and sublethal toxicological effects of component compounds (such as PAH) on exposed organisms; (3) long-term effects from persistent oil residues or recalcitrant component compounds in sheltered environments or permeable substrates; or (4) acute impacts that lead to long-term adverse impacts on population dynamics (e.g., a small spill at a sensitive breeding location).

In addition to the acute and long-term effects associated with spills, there has been growing awareness in the ecological community that the multitude of small but frequent releases of hydrocarbons (e.g., urban runoff, permitted discharges from treatment plants, vessel operational discharges) to the environment results in a significant load of potentially toxic compounds (albeit at sublethal concentrations; National Research Council, 1985). Although spills may result in spatially and temporally restricted areas of high concentration of potentially toxic com pounds, such releases generally account for a very small percentage of the total load delivered to the environment. Conversely, if the background concentration of toxic compounds attributable to chronic loads is high in a given area, small spills that would normally dilute rapidly to below levels of concern may raise the ambient background concentration above a concentration of concern for a significant length of time. Thus, the physical, acute, long-term, and/or cumulative effects of a release of floating oil are complex to evaluate. Ecological effects associated with spills of emulsified fuels can be discussed in a similar fashion, and their effects will differ based on the physical, chemical, and biological behavior of the component compounds and the setting in which the spill occurs.

UNDERSTANDING TOXICITY

The physical effects associated with spills of petroleum products (e.g., smothering or coating of floating oil on birds, mammals, or other organisms) are well documented (Boesch and Rabalais, 1987; National Research Council, 1985; Wells et al., 1995). However, toxicology studies conducted to date on the enormous number of individual hydrocarbon compounds and mixtures (such as crude oils, distillate products, or emulsified fuels) are more complex. Thus, a basic introduction to the approaches used to understand the chemistry of these complex mixtures as well as the acute (lethal) and sublethal toxicological effects seems appropriate.

The primary chemicals of initial concern from a spill of Orimulsion or other fossil fuel will be the toxic PAH. Most PAH will be released rapidly because these compounds are already dissolved in the water phase (30 percent) of the product. Although no direct measurement of the concentration of soluble PAH in Orimulsion has been made, estimates are that only about 15 to 30 µg/L (parts per billion [ppb]) would be present in the water phase of the product (French, 2000). Analyses by Ostazeski et al. (1998a, 1998b) of a nominal concentration of 3,550 mg/L Orimulsion in seawater found only 1 ppb of PAH in the filtered (soluble) phase and 16 ppb when this same amount was added to fresh water. The estimates and available data discussed in Chapter 2 suggest that the concentrations of PAH in the water during bioassays and in the field during a spill could never exceed 30 ppb. It is true that additional slow dissolution of PAH from the bitumen particles will contribute to potential biological effects, but the extent of dissolution is a function of the dilution with water, limiting the maximum concentration to 15-30 ppb. These chemical and physical factors will result in continuous exposure, since the PAH in the water are replenished by PAH from the bitumen, but the total concentration to which organisms would be exposed should never be higher than about 30 ppb.

Standard procedures with reference oils were developed to extract the water-soluble components from crude and fuel oils and to conduct tests on a range of tolerant and sensitive species. These tests were designed to measure both acute and chronic effects, as well as bioaccumulation of petroleum hydrocarbons. From these investigations in the 1970s, it became apparent that the toxic chemicals from oils were the monoaromatics (BTEX), the two-ring aromatics (e.g., naphthalenes) and the three-ring compounds (e.g., phenanthrenes). Laboratory and field studies also showed that naphthalenes, phenanthrenes, fluorenes, and dibenzothiophenes accumulated in tissues when animals were exposed to water extracts of oils. Toxicity testing with pure compounds showed that mortality did occur when animals were exposed to BETX compounds. However, these compounds are volatile and evaporate rapidly, so a constant supply would be needed for mortality to occur. The two- and three-ring compounds were more toxic and were retained in solution and tissues longer. High-molecular-weight compounds were generally low in solubility and would not reach concentrations in the water that could produce acute toxicity during the normal four days (96 hours) of exposure. However, several of these four- to six-ring compounds, such as the six-ring benzo[a]pyrene, are carcinogenic and can produce chronic toxicity from long-term exposure in sediments. Long-term effects may involve genetic damage, resulting in reproductive failure, or histological damage as observed in some fish species. More recently Ankley et al. (1995) found photoenhancement of toxicity when some of the PAH are present in the tissues of organisms exposed to ultraviolet light.

In addition to PAH that are measured routinely in water, sediment, and tissue samples and that have been investigated for their toxic effects on organisms, it should be recognized that any weathered fossil fuel may contain other components capable of producing toxicity. Recently Roland et al. (2001) described effects on marine mussels from what is generally referred to as an unresolved complex mixture (UCM). The monoaromatic UCM produced sublethal effects on the mussels, and mussels collected from a spill site contained levels of mono-, di-, and tri-aromatic UCM hydrocarbons.

Toxicity studies with oils and petroleum compounds begin with acute toxicity tests. In these tests, organisms are exposed to high concentrations of the compound of interest for 96 hours. The results are expressed as the concentration that produces 50 percent mortality in 96 hours (96-h LC_{50}). Less toxic compounds will have higher LC_{50} values, since it will take higher levels to produce 50 percent mortality. The LC_{50} values for the water-soluble fractions of reference oils and specific hydrocarbons, using a range of animals, provided the early evidence about which specific components of oils were responsible for the majority of the toxicity observed. Once data were obtained on mortality, sublethal and chronic toxicity testing was conducted to observe the magnitude of difference between the results of acute tests and chronic exposure parameters (growth, reproduction, physiology).

These tests serve as a guide for toxicity associated with specific compounds. However, aquatic toxicology has been criticized in recent decades for embracing the intellectual framework associated with toxicity bioassays to evaluate the safety of potentially harmful chemicals. Bioassays were not originally developed to determine safe exposure levels. Toxicant concentrations that fail to cause substantial mortality within a few days of exposure might nonetheless be sufficient to seriously harm populations. These problems are extremely difficult to assess with bioassays for mixtures of compounds (Box 3.1). **Further studies employ-**

Box 3.1

Weaknesses in the Use of Standard Bioassay Conditions in Assessing the Toxicity of Complex Mixtures, such as Orimulsion-400 and Fuel Oils

Standard bioassay procedures as described in American Society for Testing and Materials (ASTM) or EPA documents are most appropriate for testing single water-soluble chemicals or effluents with low particulate content (about 30 ppm [parts per million] or less). Mixtures of petroleum hydrocarbons are particularly difficult to use in bioassays. This is because there are both soluble and relatively insoluble phases, with some of the components floating in suspension and some sinking. In addition, the composition of this mixture in solution is constantly changing. Recent bioassay procedures for testing the toxicity of sediments were not designed to test sediments spiked with mixtures such as Orimulsion- 400 and fuel oils, because one portion of these mixtures will be soluble, and then volatile, whereas another portion will remain coating the substrate. Bioassay organisms that do not come in contact with or ingest floating or suspended oil droplets are exposed primarily to the soluble components. Anderson et al. (1974a,b) demonstrated that within 24 hours about 50 to 80 percent of the soluble and toxic aromatic hydrocarbons from crude and fuel oil in aerated bioassay containers had volatilized. They also showed that there was little or no change in the LC_{50} from 24 to 96 hours. Therefore, the use of 96-hour bioassays to compare the toxicity of different mixtures of petroleum hydrocarbons is not very appropriate.

Probably the best method of assessing the toxicity of complex mixtures, including dispersed oils, is a constant flowing exposure system where the exposure concentration and time to 50 percent mortality are measured (Anderson et al., 1981, 1984, 1987). Capturing and treating the waste from such testing make the use of this system difficult and expensive. The next best alternative is to expose organisms for 24 hours to known concentrations of the soluble and toxic hydrocarbons derived from the test products (water-soluble fractions of fuel oils and filtered fraction of Orimulsion-400). Bioaccumulation studies have shown that the toxic naphthalenes and phenanthrenes are accumulated by exposed organisms (Anderson et al., 1974b), which helps to explain the observed toxicity. Insufficient data are available on the bioaccumulation of these compounds or other hydrocarbons from the soluble or particulate phase of Orimulsion-400.

Given all of the problems associated with characterization of the conditions that result in significant toxicity (time and concentration of components), it is difficult to compare the results of toxicity tests with one product, such as Orimulsion-400, to those conducted on fuel oils presently transported and used in the United States. This comparison is further complicated by the fact that the majority of the PAH (including the toxic naphthalenes and phenanthrenes) contained in Orimulsion-400 are in 1- to 80-µm particles, while these compounds are initially present in the floating fuel oils and then partition into the water phase during a bioassay. A weakness in many of the toxicity study comparisons used to produce data on both No. 6 fuel oil and Orimulsion-400 is that the concentrations of some of the most toxic components (naphthalenes and phenanthrenes) were not measured at the beginning and end of the exposures. It is possible to estimate these concentrations, based on other studies (Ostazeski et al., 1998a, 1998b), but the accuracy of these estimates and the duration of exposure to these compounds are in doubt.

ing methods specifically designed to evaluate chronic or sub-lethal effects from exposure to Orimulsion, and its components, should be carried out.

With the previous discussion in mind, it is still possible to draw some general conclusions about potential toxicological effects associated with spills of Orimulsion. Available published toxicological studies examine the effects of two formulations of Orimulsion: Orimulsion-100, discontinued in 1998, and Orimulsion-400. Conclusions drawn in this report are applicable to either formulation unless specifically stated otherwise.

The soluble PAH present in Orimulsion occur at very low concentrations, compared to other heavy fuel oils and crude oils (Anderson et al., 1974a; Table 2.2). Even from spills of large amounts of crude oil, the water column seldom contains a high enough concentration for a sufficient period of time to produce acute toxicity. The concentration of toxic hydrocarbons entering the water during a spill of Fuel Oil No. 6 would be approximately 10 times greater than that introduced by a spill of Orimulsion-400. It can therefore be assumed, that under identical conditions, the concentration of soluble PAH with known acute and long-term toxicity (naphthalene; two-ring aromatics and phenanthrene; three-ring compounds) would also be lower than from a spill of an equal volume of Fuel Oil No. 6 (Anderson et al., 1974b; Neff and Anderson, 1982). Still, an examination of the potential concentration of compounds of concern from a spill of a given volume at a specific location and time is necessary to fully evaluate the environmental risks. Furthermore, many of the same chemicals that are toxic are also those that would be present in the water column. Bioaccumulation from contaminated sediments, or sediments mixed with bitumen droplets, would involve high-molecular-weight compounds (Meador et al., 1995). The very soluble alkylbenzenes (BTEX) are nearly absent from Orimulsion, so no effects from these chemicals are likely.

Consideration should be given to those compounds added to bitumen to disperse and stabilize the emulsion. Intan 400 AE constitutes 0.13 percent of Orimulsion-400. Tank studies by Ostazeski et al. (1998a, 1998b) showed that the addition of 14.2 liters of Orimulsion-400 to 4,000 liters of water (nominal concentration of 3,550 ppm) produced an AE concentration of about 3 ppm within 30 minutes and this level was maintained in solution in both the seawater and the freshwater tanks for the entire 168 hours of testing. The fact that AE is present in the water phase of Orimulsion at concentrations 100 times higher than soluble PAH added to water during bioassay studies strongly suggests that the observed toxicity in the many bioassays was primarily from AE, with some contribution from PAH. For example, if a nominal concentration of 3,550 ppm Orimulsion was found to be the LC_{50} in a given test, the water would contain about 3 ppm of AE but at the most 30 ppb of PAH. Based on studies with both classes of chemicals, the AE would be at lethal concentration (Table 3.1) while the PAH would be somewhat less than the lethal concentration.

TABLE 3.1 Acute Effects of Linear and Branched Alcohol Ethoxylate Concentrations (mg/L)

AE	Algae	Crustaceans	Fish
Linear AE	*Selanastrum cornutum* (96-h growth)	*Daphnia magna* (48-h survival)	*Pimephales promelas* (96-h survival)
n-$C_{12-15}AE_7$	1.3^b		
n-$C_{13-15}AE_7$	0.9^b	0.6^b	1.5^b
n-pri-$C_{12-15}AE_9$	0.7^b	1.6^c	1.2^c
		4.0 (LOEC, in 7-d survival tests)^b	2.0 (LOEC, in 7-d survival tests)^b
Branched AE			
$C_{12}AE_7$ (Exxal 12-based)	38.7^d	6.8^d	6.8^d
$C_{13}AE_7$ (Exxal 14-based)	37.2^d	5.9^d	4.4^d
$C_{13}AE_7$ (two methyl branches)	7.5^b	8.6^b	4.5^b
		4.0 (LOEC, in 7-d survival tests)^b	4.0 (LOEC, in 7-d survival tests)^b
$C_{13}AE_7$ (four methyl branches)	10.0^b	9.4^b	6.1^b
		6.0 (LOEC, in 7-d survival tests)^b	2.0 (LOEC, in 7-d survival tests)^b
	Skeletonema costatum (48-h growth)	*Acartia tonsa* (48-h survival)	*Scophthalmus maximus* (48-h survival)
Orimulsion-400 $C_{12-13}AE_{9-22}$ (GENAPOL X 159)		2–4	
Orimulsion-400 (fuel, per se)	500 (LOEC)^a	100 (LOEC)^a	200 (LOEC)^a

SOURCE: ^{a}Bjornestad et al., 1998; ^{b}Dorn et al., 1993; ^{c}Kravetz et al., 1991; ^{d}Markarian et al., 1989.

The second chemical used in stabilizing the emulsion, MEA, is far less toxic than AE surfactant. Most LC_{50} values for MEA (Table 3.2) are in the hundreds of parts per million of Orimulsion or two orders of magnitude higher than the soluble PAH concentration.

Computer modeling (discussed in Chapter 2) supports the view that AE contained in the water phase of Orimulsion would be diluted rapidly in open water. The available data suggest that the LC_{50} for either 48- or 96-hour exposure may be about 1 ppm, and a seven-day chronic test on the reproduction of *Ceriodaphnia dubia* reported a No-Observable-Effect Concentration (NOEC) of 0.17 ppm (Harwell and Johnson, 2000). The AE concentration would be at the acute lethal

TABLE 3.2 Toxicity of Monoethanolamine

Species	End Point	Value (mg/L)
Water flea *(Daphnia magna)*	LC_{50}, 24 h	140
	LOEL (acute toxicity)	100
Mosquitofish *(Gambusia affinis)*	LC_{50}, 96 h	337.5
Bluegill *(Lepomis macrochirus)*	LC_{50}, 96 h	329.2
Golden orfe *(Leuciscus idus)*	LC_{50}	224, 525
Goldfish *(Carassius auratus)*	LC_{50}, 96 h	170
	LC_{50}, 96 h (pH 10.1)	>5,000
	LC_{50}, 24 h (pH 10.1)	190
Rainbow trout *(Oncorhynchus mykiss)*	LC_{50}, 96 h	150
Fathead minnow *(Pimephales promelas)*	LC_{50}, 96 h	2,100
Zebrafish embryos *(Brachydanio rerio)*	LC_{50}, 8-cell stage until hatching	3,600
	NOEL, 8-cell stage until hatching	1,200
Clawed toad *(Xenopus laevis)*	LC_{50}, 48 h	220
Blue-green algae *(Anacystis aeruginosa)*	Population growth (threshold cell multiplication inhibition test)	7.5
Blue-green algae *(Microcystis aeruginosa)*	Population growth (decrease in cell multiplication), 8 d	2.1
Green algae *(Scenedesmus quadricauda)*	Population growth (threshold cell multiplication inhibition test)	0.970
Green algae *(Scenedesmus quadricauda)*	Population growth (3% decrease in extinction coefficient), 7 d	0.750
Flagellate euglenoid *(Entosiphon sulcatum)*	Population growth (>5% decrease in growth), 72 h	300
Cryptomonad *(Chilomonas paramecium)*	Population growth (5% decrease in cell count), 48 h	733

NOTE: NOEL = no-observed-effect level.
SOURCE: Davis and Carpenter, 1997.

concentration (1 ppm) after a thousandfold dilution and below the acute sublethal concentration after ten-thousandfold dilution. For most spills the water component of Orimulsion is expected to be diluted by these ratios within hours. Thus, acute lethal and sublethal effects from AE in even large spills of Orimulsion-400 are not likely to be significant (long-term effects are discussed later).

COMPARING TOXICITY VALUES AMONG DIFFERENT FUEL TYPES

In an effort to place the risk associated with spills of Orimulsion into some context, much of the existing literature makes comparisons among Orimulsion and other petroleum products such as Fuel Oil No. 6 (Table 3.3). Two factors must be accounted for in order to effectively compare available information on

TABLE 3.3 Concentrations of Compounds of Known Toxicity in Orimulsion-400 and No. 6 Fuel Oil in mg/L of Whole Product

Compounds	No. 6 fuel oil	Orimulsion-400
Naphthalenes	46,649	474
Fluorenes	12,767	248
Phenanthrenes	48,927	854
Fluoranthrenes	35,272	191

SOURCE: Ostazeski et al., 1998a, 1998b; Stout, 1999; and American Petroleum Institute, 1999.

the toxicity of a spill of a multiphase fuel such as Orimulsion-400 to other fuels. First, unlike typical fuel oils where the distribution of various compounds of concern is essentially uniform, the distribution of compounds in multiphase fuels may vary dramatically among the components. For example, in any volume of Orimulsion-400, the bulk of the total PAH (3,000 ppm; Table 2.1) is tied up in the bitumen droplets, and the extent and rate of transfer of toxic PAH from these particles to the water and organisms are not well understood. The more dispersed the droplets are, the greater will be the exchange of PAH to the water, but this will also mean high dilution in the water, decreasing the possibility of effects. The second major difficulty in comparing toxicity data for multiphase fuels with those derived for typical fuel oils is due to the behavior of the additives in water. Each component of a multiphase fuel may behave differently. For example, the bitumen phase of Orimulsion-400 demonstrates unique physical behavior that varies with salinity, etc., while the water phase simply disperses into the water column (see Chapter 2).

Together, these two factors make it difficult to compare the LC_{50} values derived for Orimulsion-400 with available data derived for floating oils such as No. 6 fuel oil. For example, toxicity tests intended to evaluate the risk to organisms in the water column from spills of floating oils typically focus on the toxicity of the WSF or water- accommodated fraction (WAF) below the oil slick. Johnson et al. (1998), however, expressed the results of bioassay toxicity tests in terms of the nominal concentration (i.e., amount of total added product) of Orimulsion-400. Thus, it is difficult to directly compare LC_{50} values based on the hydrocarbons in solutions (WSF) of oils with those reported in terms of added concentration of the whole product of Orimulsion-400.

Based on numerous evaluations (French, 2000) and tank tests (Ostazeski et al., 1998a, 1998b) it can be estimated that up to 0.001 percent (30 ppb) of the total PAH (3,000 ppm) in Orimulsion resides in the water phase. Because understanding initial concentrations is imperative to understanding the potential toxicological effects from a spill, these uncertainties greatly limit the usefulness of bioassay

TABLE 3.4 Acute Toxicity of Oil-in-Water Dispersions of Orimulsion-400 to Aquatic Organisms

Test Species	Test Type (Reference)	Orimulsion-400 (ppm nominal)
Bacteria		
Microtox	Salt water,	2,579-2,800[a]
(Vibrio fischeri)	15-min test (4,3)	
ALGAE		
Selenastrum capricornutum	Fresh water, 72-h chronic (4)	>1,000 (100% stock)
Skeletonema costatum	Salt water, 72-h chronic (1)	500
Invertebrates		
Crustacean	Salt water,	650
(Acartia tonsa)	48-h acute (1)	10 (egg production)
	22-d chronic (1)	
Sea urchin	Salt water,	
(Lytechinus pictus)	20-min fertilization (3)	5,730
(Paracentrotus lividus)	Fertilization (4)	296
Mysids	Salt water,	
(Mysidopsis bahia)	96-h acute (5,2)	24.6-42.7[a]
Rotifer	Fresh water,	
(Brachionus plicatilis)	24-h acute (4)	633
Branchiopod	Salt water, 24-h acute (4)	
(Artemia salina)		>1,000
Amphipod	Salt water,	601 (10-d)
(Corophium orientale)	10- and 20-d chronic (4)	293 (20-d)
Crustacean	Fresh water,	
(Thamnocephalus platyurus)	24-h acute (4)	>1,000 (100 % stock)
Water flea (Daphnia magna)	Fresh water, 48-h acute (4)	464
Fish		
Turbot	Salt water, 96-h acute (1)	<2,000
(Scophthalmus maximus)		
Inland silverside	Salt water, 96-h acute (5,2)	114-200[a]
(Menidia beryllina)		
Rainbow trout	Fresh water, 96-h acute (3)	301
(Onchorhynchus mykiss)		
Threespine stickleback	Salt water, 96-h acute (3,2)	3,200
(Gasterosteus aculeatus)		

[a] Two values available for this test.
SOURCE: (1) Bjornstead et al., 1998 (WAF used); (2) Calabrese et al., 1995; (3) Jokuty et al., 1995; (4) Golder Associates Geoanalysis, S.r.l, 1999; (5) Johnson, 1998.

studies completed to date. However, by keeping these relationships in mind and by assuming a maximum concentration of bioavailable PAH of 30 ppb, it is possible to examine and make some general comparisons among previous toxicity data (Table 3.4) for Orimulsion-400 and floating oils such as No. 6 fuel oil.

For the reasons expressed above, the best comparisons from Table 3.4 would be the ppm concentrations measured in the water during exposures to both No. 6 fuel oil and Orimulsion-400 that produce either 50 percent mortality in a specific number of hours, or a reduction of 50 percent in a sublethal end point. Some of the most sensitive marine organisms used in toxicity testing are mysids (*Mysidopsis*) and silversides (*Menidia*), first used to assess petroleum hydrocarbon toxicity by Anderson et al. (1974a). Johnson et al. (1998) reported the results of a 96-hour exposure of mysids to a filtered WSF of Orimulsion-400, producing a nominal LC_{50} of 7,070 ppm of Orimulsion-400. Using the factors for PAH and AE discussed above it can be estimated that the mysids were exposed to 7 ppm of AE and 30 ppb of PAH. Both of these concentrations are near the LC_{50} values for these chemical classes, so it is only possible to state that the combination produced the observed effects. These investigators also reported an LC_{50} for *Menidia* exposed to the filtered WSF from a nominal concentration of Orimulsion of 12,800 ppm. Using the same factors, this exposure would be to 12 ppm of AE and 30 ppb of PAH, and there is certainly evidence from AE testing (Table 3.1) that 12 ppm of AE would be above the LC_{50} value. There are many other toxicity testing data points in various tables from multiple publications, but without specific chemical analysis of PAH in the water, the findings cannot be readily compared to tests with other fuel types. It is clear that the maximum concentration of PAH that can be produced in the water during either a toxicity test or a spill of Orimulsion is less than that produced by floating oils. It is important, however, to consider the additional contribution to toxicity from the AE in this product, as initial concentrations can be toxic to exposed organisms.

To measure the bioavailability and effects of PAH from Orimulsion formulations at low concentrations (0.01-1.0 mg/L), sea scallops were exposed for 60 days (Armsworthy et al., 1999). No impacts were found with respect to survivorship or somatic and reproductive tissue weight, but some increases in clearance rate and absorption efficiency were noted. Analysis of the tissues after the 60-day exposures found the highest concentrations in the digestive glands (30 ppm) and gonads (12 ppm). The accumulation in the gonads after 60 days represented a factor of 100, which is the same magnitude reported for PAH in other studies. The only other study to measure uptake of PAH from Orimulsion was with oysters, but the detection limits for the analyses were so high that the data are not useful. Therefore, the potential for uptake is known for only one test organism, but the findings agree with other studies of bivalves using specific hydrocarbons or extracts of oils. It should be recognized that many of these earlier studies also followed the duration of PAH contamination in the organisms after exposure was terminated, and found clearance of the tissues by metabolic activity, depuration, or both.

In the only study that examined the effects of dispersed Orimulsion on wildlife, mallard ducks exposed to 100 ppm Orimulsion-400 in a freshwater holding pond exhibited body weight and histopathology similar to controls (Wolfe et al.,

2001). Coverage of waterproof plumage was noted to be less in treatment birds in that bitumen residue adhered to 40-45 percent of the bird's plumage, most where the bird's body contacted the water. After three days of preening and bathing in clean water, half of that amount remained, indicating that exposure of these birds is likely by ingestion. Although this study found the effects of Orimulsion exposure to be less injurious than those of a heavy fuel oil there was some loss of waterproofing in plumage at a low Orimulsion concentration and four hour exposure duration.

Bivalves exposed to PAH from Orimulsion, or from any other type of fuel spill, have the potential to be ingested by mammalian or avian predators. This exposure route to higher trophic levels is not well studied but should be acknowledged (Duffy et al., 1994; Fry and Lowenstine, 1985; Leighton et al., 1983). Although vertebrates have a high capacity for metabolizing aromatic hydrocarbons including PAH (Spies et al., 1996), there is a need for studies addressing the possible reproductive effects of ingestion on predators and other effects on higher trophic levels.

Data on ecological effects of the dispersed form of Orimulsion on vascular plants comes from work with Orimulsion-100, which has a similar bitumen component but different additives (surfactant and magnesium) and viscosity than Orimulsion-400. Given the similarities of the bitumen component of the two formulations, the results can be used to assess the toxicity of exposure of these plants to the droplets and PAH. Data indicate that long-term exposure (21 days) to a water column concentration of 10,000 ppm Orimulsion did not result in mortality to the seagrass *Thalassia testudinum* (Ault et al., 1995). However, lower concentrations (1,000 ppm) resulted in increased leaf senescence (Ault et al., 1995). Because whole product Orimulsion (not just the bitumen component) was used in this study, it is possible that the surfactant and magnesium components contributed to this effect.

Although *Thalassia* exhibited a negative response to dispersed Orimulsion, it continued to grow and produce new leaves. This resiliency may be due to shoot and root meristems that are buried in sediment and protected from toxic substances in the water column (Ault et al., 1995). Continued growth and refoliation are also supported by nutrients stored in buried rhizomes.

In addition to PAH in the water phase of Orimulsion-400, there is 0.1350 percent (1,350 ppm) AE in the neat fuel. Based on tank testing and simple dilution calculations, only the initial pulse of these chemicals into a closed system would produce concentrations in the sub-lethal effects range for algae (Table 3.3).

Acute toxicities to fish of both branched and linear AE generally varies from <1 to 10 ppm (Table 3.1). Although the data from linear AE are more extensive than those for branched (Talmage, 1994), it appears that the carp and minnow families are more tolerant than commercial, recreational, and saltwater fish, such as sunfish, trout, or cod. Algae showed little or no observable effects from exposure to concentrations of one family of branched AE that were acutely toxic to fish and invertebrates (Markarian et al., 1989). Yet branched AE used in experiments by Dorn et al. (1993) were comparably acutely toxic (i.e., in the 4.5 to 10 mg/L range), with the freshwater fathead minnow being the most sensitive and a green alga the least sensitive. However the criterion for the algae was growth, while that for the fish was survival. Acute and chronic toxicities of nominal concentrations of specific surfactants provide only approximations of the risks to the specific type of organism used in the test, and not to the health of the waterway.

Branched AE are generally less acutely toxic than linear AE (Table 3.1) and are more resistant to biodegradation so they can be more persistent in the environment. Talmage (1994), citing works of Dorn et al. (1993), Kravetz et al. (1991), and Markarian et al. (1989), found linear AE to be up to forty-fold more acutely toxic than branched AE to the growth of a common green planktonic alga, and from two- to four-fold more toxic to a freshwater crustacean zooplankter and a common minnow (Table 3.1). In seven-day toxicity studies, Dorn et al. (1993) found that the lowest-observed-effect concentration (LOEC) of branched AE on the survival of the water flea was the same as, or 50% less toxic (depending on the number of branches) than linear AE, and branched AE were equally or two-fold less toxic for the minnow. Bjornestad et al. (1998) studied GENAPOL X 159 (the branched AE in Orimulsion-400) in survival tests with a common marine planktonic crustacean, *Acartia tonsa*, and reported an LC_{50} of 2 to 4 mg/L. These concentrations are comparable to those of linear AE and about two-fold more toxic than the branched AE used in two-day tests by Dorn et al. (1993) described above. No studies were performed on fish or algae with the surfactant, but tests with the Orimulsion-400 emulsified fuel, showed that it was up to fifty-fold less toxic to the planktonic crustacean than the surfactant and of low concern to a marine alga and a species of bottom fish.

The data for MEA in the water phase of Orimulsion-400 (Table 3.2) show it to be relatively nontoxic, since even *Daphnia* had a 24-h LC_{50} of 120 ppm. Thus, because MEA has very low toxicity, evaluation of the potential ecological effects of a spill of Orimulsion-400 should be based on the sum of toxicity and long-term effects of exposure to the PAH and AE in solution from spills in both fresh and salt water.

UNDERSTANDING ECOLOGICAL RISKS ASSOCIATED WITH SURFACTANT

Although the impacts of petroleum hydrocarbons on marine and aquatic organisms have been studied for years, the long-term chronic effects of the surfactants used in Orimulsion-400 and many other household products, are less well understood. Consequently, additional discussion of these compounds seems warranted. Surfactants in oil and bitumen emulsions are predominantly two types: (1) the AE and (2) the alkylphenyl ethoxylates (APE) and nonylphenol ethoxylates (NPE), together referred to here as APE (Karsa, 1998; Talmage, 1994). In 1998 an estimated 465 million pounds of AE and LAS were used predominantly as surfactants in household products in the United States (Stanford Research Institute, 1998). Given their widespread use, their possible toxicological interactions (i.e., additive effects), their presumed concentrations in some major U.S. rivers, and the risks they pose in aquatic systems, major synthetic surfactants (AE, APE, and LAS) are receiving greater attention from environmental protection agencies worldwide. This is especially true since background concentrations of surfactants and their intermediary decay products, together with their additive effects with indigenous pollutants, may already have impacted aquatic resources and human health, notwithstanding the effects from accidental spills of emulsified fuels.

Criteria for allowable concentrations of the synthetic commercial blends of surfactants (AE, APE, and LAS) apparently do not exist in the United States. Yet their reportedly low concentrations in the 1980s may only reflect dilution by the very high discharges in U.S. rivers, compared to the smaller rivers in other industrialized nations (Ahel et al., 2000; Naylor et al., 1992). Currently, there is no monitoring program in the United States designed to track changes in loading rates of specific surfactants to rivers. Only one study in the late 1980s reported a concentration of 0.5 ppb above, and 1.1 ppb below, an effluent of the commonly used AE in a river (Talmage, 1994).

In view of this, it is instructive to view environmental risks of spilled emulsified fuels from another perspective. A spill of the entire contents of one barge loaded with 3,000 barrels of Orimulsion-400, at 6.2 barrels per metric ton, would release about 1,380 pounds of AE (at 0.13 percent by volume in Orimulsion-400). Were this to occur in the lower Mississippi River at average discharge (about 500,000 cubic feet per second [cfs]), the concentration of surfactant would be about 80 ppb immediately downstream of the spill, if the background levels were zero. This concentration (80 ppb) is about one-fourth of the 0.28 ppm concentration derived for use in risk assessment of AE, estimated from stream mesocosm investigations (Dorn et al., 1997). However, it is near the predicted NOEC (110 ppb) for AE, estimated to be 50 to 100 times lower than the predicted environmental concentrations, derived from monitoring results of removal rates for AE and other surfactants by sewage treatment facilities (van de Plassche et al., 1999).

SPILL SCENARIOS

In an effort to understand the ecological risks associated with spills of Orimulsion in a variety of environments, the potential physical, acute lethal and sublethal effects that may occur in the six spill scenarios are discussed below (based on a review of existing literature). For ecological risk assessment, the six scenarios can be broadly combined into three major groups: (1) marine and brackish (salinities greater than 5-7 psu); (2) fresh water (salinities less than 5 psu); and (3) on land, including wetlands. The discussion for these groupings begins with a general statement about factors contributing to overall effect, followed by a greater discussion of the potential effects on four major groups of organisms (i.e. wildlife, including birds and mammals; water column resources, including fish and invertebrates; benthic organisms, including epifauna and infauna; and nonplanktonic primary producers or vascular plants, including kelp in marine settings and plants and trees in freshwater settings).

Marine and Brackish Water Environments

As discussed in Chapter 2, some of the dispersed bitumen droplets will coalesce and surface as either tarballs or tar patties. The amount of re-floating bitumen will be a function of spill volume, water density, and current dynamics and is difficult to predict. It could range from zero to as much as one-third of the spilled volume. Floating bitumen tarballs will be very sticky and persist for a long time. Recovery of floating bitumen will vary by spill volume, distance from shore, and oceanographic conditions that would either disperse or concentrate the tarballs.

Effects on Wildlife

Floating bitumen is likely to behave much like pelagic tar. Therefore, the primary concern for wildlife will be contact with any sticky tarballs on the water surface or those that strand on shorelines. Even small spills of a persistent oil can impact hundreds to thousands of marine birds, such as the 600-barrel spill of a heavy crude oil off California that killed an estimated 9,000 marine birds (Page and Carter, 1986). Birds that spend most of their time on the water surface in dense flocks (e.g. seabirds, waterfowl) are at greatest risk. The effects of pelagic tar on sea turtles have also been well documented (Witham, 1983). Turtles feed on objects floating at the water surface, and therefore are susceptible to ingestion of tar balls that can block the oral cavity and digestive tract. Floating tar can coat the flippers and the mouth can become coated as the turtle attempts to clean its flippers.

Tarballs stranding on the shoreline will be very sticky and adhere to the intertidal substrate. The degree of coating will depend on the volume spilled, the distance from shore, and the energy of the wind moving tarballs toward shore. Complete coating is unlikely, but patches of shoreline may contain a sufficient quantity of the emulsion to impact infauna and epifauna.

Effects on Water Column Resources

The bioavailability of PAH bound in the particles has not been determined. This is a factor to be considered in light of the filter feeding of either benthic or pelagic organisms that pass the particles over their membranes or ingest them. The most sensitive filter feeders in the water column would likely be small fish, whose gills could become clogged, or that could transfer PAH to the circulatory system and accumulate it in the liver. By "coughing," fish can dislodge particles from the gills. Uptake of PAH from suspended bitumen droplets in the water column by animals is likely to be limited because of low uptake efficiencies of PAH, reflecting slow kinetics and short residence time in the gut (Meador et al., 1995).

Thus, the major toxicological threat is posed by PAH already dissolved in the water fraction of Orimulsion and subsequently dissolved from the droplets when diluted. Modeling results for a 10,000-barrel spill of Orimulsion in Delaware Bay (French McCay and Galagan, 2001) indicated that total PAH levels dropped below 25 ppb (acute toxicity level for the most sensitive species exposed over an extended period) within about 12 hours. In areas where the concentration of dissolved PAH is rapidly diluted, acute toxicity impacts on water column resources are expected to be low. Impacts would be higher for conditions in which dilution is slower, such as water bodies that are confined, have low flushing rates, or are very shallow.

Effects on Benthic Organisms

Spills of Orimulsion are not expected to affect benthic organisms in deep water because the droplets tend to suspend in the top 3 meters, and any eventual sinking due to adsorption onto natural particles will be slow and widely dispersed. In shallow nearshore and estuarine settings, benthic organisms are potentially at risk from three pathways of exposure. First, the dissolved fraction could have acutely toxic impacts. However, because of rapid dilution with depth benthic organisms would typically be at low risk. Second, suspended bitumen droplets and bitumen-sediment agglomerates could be mixed throughout the water column, where they could be ingested by benthic filter feeders. Sessile organisms will be the most at risk. PAH from Orimulsion can accumulate in shellfish tissues (Armsworthy et al., 1999) resulting in potential transfer to higher trophic levels. Food web magnification of PAH is unlikely however, because organisms at the higher trophic levels often have the greatest capacity for PAH metabolism (Meador et al., 1995).

The third pathway of exposure is via tarballs that sink after mixing with sediment, either in the surf zone or after stranding onshore. This sticky oil can coat epifauna as it rolls around on the bottom, especially where the oil accumulates in a trough or pit. It is difficult to estimate how much of an Orimulsion spill would behave this way, but similar behavior has been observed for spills of heavy oil (National Research Council, 1999).

Effects on Nonplanktonic Primary Producers

Macrophytic algae (e.g. kelps, "seaweeds") and vascular plants (seagrasses, mangroves, salt marsh grasses) are important components in coastal intertidal and subtidal communities. These organisms fix carbon and may contribute significantly to nearshore food webs. They provide vertical and horizontal structure and a substrate for attachment of other organisms, they function as breeding and nursery areas for many other ecologically important organisms, and they stabilize intertidal and subtidal sediment.

Orimulsion may interact with coastal nonplanktonic primary producers either as a sticky floating bitumen residue that is predominantly influenced by wind or as a dispersed, nonsticky fluid moving with currents and tide. Clumps of floating Orimulsion residue reaching the shore may come in contact with emergent vegetation of the intertidal zone (e.g. salt marsh grasses, mangroves), adhering to plant tissue and forming patches of oiled vegetation. Associated epiphytic organisms within these patches will likely be smothered. There is no information in the literature regarding the potential effects of Orimulsion floating residue on vascular plants, but it is likely that they are similar to those of other weathered heavy fuels (Michel et al., 1995).

The dispersed form of Orimulsion may interact with subtidal as well as intertidal primary producers. A study by Ault et al. (1995) suggests that, at least for the seagrass *Thalassia*, >1,000-fold dilution of Orimulsion (which would occur rapidly under most spill scenarios) would not result in long-term injury to the vegetation from water column exposure. The reported sublethal effects at lower concentrations, however, suggest a potential for short-term stress in affected *Thalassia* populations, which may in turn impact other components of the seagrass community. Potential effects on other species of seagrass are unknown. Chronic effects on the entire community from the accumulation of bitumen bound to the sediments are also uncertain.

Because *Thalassia* has a large proportion of its living biomass buried in the sediment, it is likely to survive a coastal spill of Orimulsion. This suggests that other submerged or emergent vascular plants with high proportions of buried biomass (e.g. *Spartina*) could survive such a spill as well. However, sublethal impacts, such as leaf chlorosis or death, will affect plant production and may have repercussions throughout the food web.

Another important finding from the *Thalassia* work was the decrease in water column oxygen concentrations in treatments with Orimulsion added (Ault et al., 1995). This effect may have contributed to the invertebrate mortalities observed and should be considered as a potential risk to these organisms in estuaries where dissolved oxygen levels are already low.

Freshwater Environments

There are two important differences in the behavior of Orimulsion spills in fresh water versus seawater: (1) because Orimulsion is denser than fresh water, spills will result in dispersal of the bitumen droplets into the water column with no surfacing; and (2) the surfactant and stabilizer associated with Orimulsion are more stable in fresh water, thus there is less tendency to form clumps. Therefore, spills in flowing fresh water such as rivers and streams have a greater tendency to remain completely distributed within the water column, with no surfacing or settling. There will still be some attachment of droplets to natural material, but less than in estuarine or marine conditions, because of the stability of the dispersant. As the suspended droplets move down stream, high concentration pockets of Orimulsion may build up on the bottom where there are back eddies or depressions. If such an accumulation were to take place, there could be impacts on fish or invertebrates spending significant time at these sites. With an increase in flow, these pockets would be flushed downriver.

For spills into quiescent water bodies, the bitumen droplets are expected to form a thick cloud that settles through the water column. Based on tank tests, the droplets would remain dissociated from one another and slowly settle to the bottom. Because the droplets are so small, they would readily be resuspended by any disturbance. The soluble components (PAH, surfactants, stabilizers) would diffuse into the water column above the settling droplets and subsequently move into the surrounding water, slowly diluting.

Effects on Wildlife

Risks to wildlife in the water column swimming through the dispersed form of Orimulsion-400 are unknown. There is no information on whether bitumen droplets suspended in the water column will stick to fur or feathers. Swimming at the surface of a water body containing dispersed Orimulsion does present a risk to wildlife, however, because coalesced floating bitumen has more of a tendency to stick to fur or feathers. Mallard ducks exposed to completely dispersed (100 ppm) Orimulsion in a pond had bitumen droplets stuck on their feathers, resulting in a reduction in water proofing (Wolfe et al., 2001). It is possible that the agitation of the water column by the swimming of the ducks in this confined water body promoted the coalescence and floating of the bitumen droplets.

Effects on Water Column Resources

Because all of the product mixes into the water column, spills of Orimulsion into freshwater environments have a greater potential to impact water column resources than floating oil spills. The extent of impact is directly related to the spill volume and rate of dilution of the PAH and AE associated with the water phase and the PAH desorbed from the dispersed droplets. Dilution of AE by a factor of 1,000-10,000 would bring concentrations below levels of concern for short-term exposures (1,350 mg/L in neat Orimulsion, 1-2 mg/L as an acute LC_{50}, and 0.1 mg/L as a NOEC). The difference between the potential impacts of PAH and AE is that the bitumen droplets contain about 3,000 ppm of PAH; thus, PAH desorption from bitumen droplets after the emulsion is broken provides additional soluble PAH to the water column. The rate and extent of PAH exchange from the bitumen to the water is not well understood, but will be a function of dilution in the water. However, PAH concentrations will not exceed 30 ppb, the estimated initial concentration in the water phase of the fuel. Depending on flow rates, spills into rivers may have a slower rate of dilution than spills into bays and oceans. In rivers, motile species may be able to escape as the plume of droplets and dissolved PAH moves downstream, but nonmotile species could suffer acute toxicity from larger spills.

Effects on Benthic Organisms

Spills of Orimulsion in freshwater environments have a greater potential than floating oil spills to impact benthic organisms because bitumen droplets are denser than fresh water. Very small bitumen droplets will remain in suspension under all but the most quiescent settings, so filter-feeding benthic organisms could be exposed to the cloud of droplets as they move past. Based on limited data for sea scallops (Armsworthy et al., 1999), PAH uptake from solution will be approximately as described for oils and specific hydrocarbons (Meador et al., 1995), but accumulation from bitumen particles is not clearly understood. The risk of exposure to particles or oxygen depletion would be greatest for the most quiescent settings where bitumen could settle out of the water column and accumulate as a cloud of droplets suspended above the bottom. Benthic organisms would also be exposed to dissolved PAH and AE concentrations similar to those encountered by water column resources discussed above.

Effects on Nonplanktonic Primary Producers

Concentrations of bitumen associated with spills in rivers and harbors will present minimal risk to submerged macroalgae or vascular plants except in the most quiescent settings. Although macrophytes will be exposed to PAH and AE concentrations similar to those encountered by water column resources, they tend to be much less sensitive, which suggests that spills in freshwater settings pose minimal risk to these organisms.

On Land, Including Wetlands

Orimulsion spills on land are expected initially to flow downgradient like a heavy fuel oil. However, the emulsion is expected to break, allowing the water phase to drain away, leaving a surface layer of very sticky bitumen. On land and in wetlands, there may be some penetration of free product Orimulsion into coarse substrates.

Effects on Wildlife

Drying of the product on land and in wetlands would produce a very sticky substance, posing a hazard of coating for any wildlife entering the area. Small furbearers would be at risk of fouling, ingestion during cleaning, and habitat displacement. The risk to biota in the path of the spill would be primarily from smothering.

Effects on Vascular Plants

Plants may be impacted by spills of Orimulsion by smothering or through toxic effects resulting from exposure to roots or stems. Smothering occurs if large patches of Orimulsion are left in place on the ground surface, reducing oxygen diffusion into soils and interfering with root respiration. Although cleanup on land is possible (even a layer of the product in a marsh-type environment could be removed and recovered), spill cleanup may injure plants through physical damage to tissues, soil removal, or soil compaction.

Summary of Ecological Effects

The ecological effects associated with spills of petroleum can generally be categorized as (1) physical effects associated with smothering or coating (2) acute lethal and sublethal toxicological effects of component compounds (such as polycyclic aromatic compounds) on exposed organisms; (3) long-term effects from persistent oil residues in sheltered environments or permeable substrates or acute impacts that lead to long-term adverse effects on population dynamics (e.g., a small spill at a sensitive breeding location); and (4) effects associated with chronic releases. Thus, the more significant conclusions presented so far regarding ecological effects from spills of Orimulsion can be categorized in a similar fashion.

Physical Effects

One of the most damaging impacts during spills of floating oils is on birds coated by the oil when landing in or diving through the oil slick. In most instances, however, a spill of Orimulsion would not form a surface slick. In open-ocean or coastal environments (salinty >5psu) Orimulsion tends to coalesce, resurface, and form floating tarballs. These tarballs will be highly persistent and would contribute to pelagic tar in the ocean that affects both marine birds and turtles. Another key impact from oil slicks is the covering of the intertidal zone with a layer of spilled oil, some of which may move back into the subtidal zone. Complete coverage of the intertidal zone with a coating of Orimulsion is unlikely, because only a fraction of the spill is expected to re-float. **It is therefore likely that impacts from a spill of Orimulsion on birds, marine mammals, and intertidal organisms may be significantly less than spills of floating oils.**

Acute Lethal and Sublethal Effects

Floating oils, including crude oil and refined products, presently transported and used in the United States contain up to 10 times higher concentrations of PAH than Orimulsion-400. The monoaromatics (BTEX) contribute to the toxicity of spilled fuel oils in the early stages of a spill, but these are present in such low concentrations in Orimulsion that they pose little risk to aquatic organisms. In addition, most of the PAH in Orimulsion occurs in the bitumen droplets, and the bioavailability of these compounds to non-filter-feeding organisms will likely be low, since the chemicals would first have to exchange into the water, at very dilute concentrations. **The accumulation of PAH by filter feeders from the soluble phase and bitumen droplets is not well understood at present. Studies should be undertaken to fully evaluate the bioavailability of PAH to filter feeders.**

There are, however, significant problems and complexity in attempting to compare the effects of Orimulsion to those of No. 2 fuel oil or No. 6 fuel oil. Previous research has shown that the soluble components of floating oils (BTEX, naphthalenes, phenanthrenes, etc.) are responsible for the observed toxicity to sensitive aquatic organisms. Bioaccumulation of naphthalenes and phenanthrenes during exposure provides additional evidence of their relationship to the toxic effects on animals. Some organisms are capable of metabolizing or at least depurating these compounds over time if a lethal dose is not received. However, there can be long-term effects from these exposures, such as histological damage in fish.

In attempting to estimate the concentrations of toxic PAH in the water during a toxicity test with Orimulsion, it was necessary to use chemical data from tank tests without animals. **Any additional testing with Orimulsion should provide the detailed chemical analyses necessary to make this type of comparison.** Whole product testing with sensitive fish species (rainbow trout and silversides) showed that levels of about 100 to 300 ppm produced 96-h LC_{50} values. These fish were exposed to both the particulate and the soluble phases of the product, as would be the case during a spill, so these values are likely good estimates of the potential acute effects on fish from the dispersed product.

Some additional key questions remaining on the impacts of Orimulsion relate to the bioavailability of PAH from the whole Orimulsion product and from the bitumen droplets only. Short-term bioaccumulation studies for key commercial crustaceans and bivalves of the whole product dispersed in water would answer questions about potential uptake and seafood safety. **To more fully address this question, standard bioaccumulation tests should be conducted, with exposure of organisms both to the whole product mixed in sediment and to the solid phase only (after filtration to remove the 30 percent water) mixed in sediment.** These studies would have to demonstrate both solid-phase toxicity and bioavailability.

Long-Term Effects

The long-term effects of oil spills are a function of the bioavailability of the chronically toxic, high-molecular-weight hydrocarbons that remain in and on fine sediments. No spills of Orimulsion have occurred in which the potential effects of such coating or mixing with fine sediments could be observed. **Studies are recommended that measure the bioavailability of PAH from mixtures of Orimulsion and sediments, and the acute as well as chronic effects of these exposures.**

Effects from Chronic Releases of Surfactants

Surfactants are widely used, they are acutely and chronically toxic, they may interact with or add to the toxicity of other pollutants, and they are not monitored in U.S. waterways. Currently there are no criteria in the United States for allowable concentrations of surfactants or their biodegraded intermediary products. Current background concentrations of surfactants and degraded products in U.S. waterways are unknown. Only an approximation of risks from spills of emulsified fuels in U.S. waterways can be made, using models that are based mostly on toxic concentrations derived from acute toxicity studies of nominal amounts of the untreated toxicant. **Better understanding of the significance of AE and its by-products in the environment would require the establishment of monitoring programs for specific types of synthetic, commercial blends of surfactants and their more persistent biodegraded intermediary products in U.S.**

waterways. Furthermore, the establishment of criteria for allowable (low risk) concentrations of each commercial blend of surfactant and its biodegradation products would be needed to minimize environmental and health risks.

Long, linearly arranged alkyl chains (up to C_{20}) are less acutely toxic than short chains and branched chains, and AE with longer EO_n moieties are less toxic as well. Biodegraded intermediary products (PEG and carboxylated AE) apparently persist longer than die-away tests (28-day) have shown; the period and products of ultimate degradation are unknown. Therefore, surfactants with longer, less complexly branched alkyl chains and long-chain EO moieties should be used in emulsified fuels (as was done with Orimulsion). Complexly branched hydrophobic and hydrophilic moieties should be avoided in emulsified fuels because of their recalcitrance to degradation in sewage treatment facilities. The use of naturally derived surfactants in emulsified fuels should be investigated.

4

Efficacy of Response: Summary and Evaluation of Available Information

In this chapter the available information on response technologies is evaluated with respect to the likely efficacy of response in the six settings described in Chapter 2, namely, marine—open water; marine—nearshore; estuarine (brackish water); nontidal river; fresh water—quiescent; and on land near water. Most of the literature on response options for spills of emulsified fuels describes equipment and systems that are similar to those traditionally used to respond to spills but modified to have improved effectiveness for spills of emulsified fuels. However, because there have been no significant spills of emulsified fuels, the modified equipment and the strategies for their use have been untested in real cleanup situations. Therefore, no actual spill data for determining actual or measurable effectiveness are available. These response options and their modifications are evaluated to the degree possible in the following sections.

It should be noted that the effectiveness of oil response in containment and recovery of spilled oil has been historically low. In the response community, recovery of 20 percent of the spill volume is considered to be a good effort. Where dispersants or in situ burning are used effectively there is little or no additional response or recovery.

MARINE OPEN-WATER RESPONSE

In recent years, alternative technologies have been accepted as response tools to be considered as part of the oil spill response arsenal in the United States. Alternative response technologies usually include, but are not limited to, the use of dispersants, in situ burning, and bioremediation. Since some emulsified fuels, such as Orimulsion-400, already have a dispersant added in the form of their emulsifying surfactant, at the instance of spillage they disperse into the upper part of the water column. Therefore, the aerial dispersant application operations that are normally considered the first response in many areas of the United States for most spilled oils in the open water, have already taken place with emulsified fuel products.

Since 1989, BITOR and its subsidiaries, in conjunction with their primarily utility company clients and U.S. and international government agencies, have studied the response to and possible cleanup techniques for spills of Orimulsion-100 and Orimulsion-400. These studies provide a large volume of information regarding available cleanup techniques and the individual merits and failures of response for these emulsified fuel products. Table 4.1 summarizes the effectiveness of response options for spills of emulsified fuels in marine open-ocean environments. It is clear that in an open-water marine environment, the preferred response methodology would be to monitor the naturally dispersed bitumen plume using the Surveillance and Monitoring for Alternative Response Technology (SMART) protocols in place for monitoring chemically dispersed spills of other types of oil. This sampling and analysis approach could also be augmented with discrete water column sampling for separate dissolved and dispersed bitumen droplet phases (Payne et al., 1999), to provide calibration for the UV fluorescence approach of the SMART Protocols and validation of computer model predictions of dissolved- and dispersed-bitumen droplet PAH concentrations.

The availability of existing sophisticated computer three-dimensional modeling programs will support and enhance required surveillance operations (French et al., 1997). If any of the spill re-floated and was found, cleanup would be initiated using existing cleanup technology developed and available for Orimulsion, as well as technologies available for other Group V oils that are presently being transported in the United States (National Research Council, 1999).

Many of the tests for equipment specifically designed for response to spills of Orimulsion products have been carried out on the open-ocean to determine the ability of responders to recover dispersed bitumen in that environment (Hvidbak and Masciangioli, 2000). As is the case with most open-ocean oil spill response equipment, some equipment developed specifically for response to Orimulsion spills and evaluated during the documented tests, such as the Tar Hawg, the forced adhesion and floatation (FAF) system, the PNP Re-floater, and the deep-skirted containment booms, can reasonably be deployed and perform to some degree of efficiency in calm, open-ocean conditions, especially during an intentional spill demonstration scenario (Bitor America Corporation, 1997). For persons not familiar with equipment that has been developed specifically for Orimulsion, the Tar Hawg is a belt skimmer that has teeth built into the belt to enhance bitumen adhesion. The FAF and PNP Re-floater systems are two different methods of pumping both water and Orimulsion from the dispersed plume and forcing air into the pumped material, causing the bitumen to float. As the

TABLE 4.1 Effectiveness of Response Techniques for Spills of Emulsified Fuels in Marine Open-Water Environments

Response Technique	Effectiveness for Emulsified Fuels	Special Issues
Monitoring and tracking	Similar to success of tracking intentionally dispersed oils	SMART protocols should be used and augmented with large-volume water sampling systems to analyze dissolved and dispersed bitumen PAH concentrations
Mechanical re-floating and forced adhesion and flotation of dispersed emulsified oil	Even under optimum conditions, amount of oil that could be found, contained, and re-floated would be minimal	Very specialized equipment; not proven under spill conditions; little hands-on experience by responders. Logistics may be overwhelming to deploy
Long-skirted containment boom	Limited to upper 3 m of water column, difficult to be towed by boats, must be allowed to drift with current	To be effective, must be deployed before the dispersed plume spreads in the water column
Skimming	Low overall because of limited amount of bitumen that will re-float in this environment	Floating bitumen "clumps" are very sticky and viscous, requiring specialized skimmers and pumping systems
Trawl nets	Effective for containing and recovering weathered bitumen in tests. Offloading in significant incidents may be an issue for this recovery technique	Smaller mesh nets may be more effective but will increase drag for towing vessels. More testing required
Pumping	Moving recovered bitumen from skimmers to storage and offloading storage devices will require heavy oil pumping systems and perhaps heating systems	Successful tests have been conducted on small amounts of spilled Orimulsion. However, the effectiveness and efficiency of tested devices may differ greatly for large uncontained spills
On-water storage	Disposable bladders or bags and tanks barges may be used	Tank barges and tanks on skimming vessels, should have heating coils installed to facilitate offloading
Disposal of recovered weathered bitumen	Same as for other crude oil and oil products	Shoreside recycling and incineration are probably best options available

name suggests, the deep-skirted containment boom is a normal offshore-type oil spill containment boom with a 9- to 12-foot skirt. However, although responding to locate and recover any floating weathered bitumen is practical and reasonable, the notion that requiring an operation to be mounted with the goal of tracking, re-floating, and then recovering a dispersed oil of any type in the open ocean would be unrealistic and unreasonable.

The disposal of recovered weathered bitumen poses the same problems as the disposal of other recovered crude oil and oil products that are spilled (Bitor America Corporation, 1999).

MARINE NEARSHORE RESPONSE

In the nearshore marine environment, the surveillance and monitoring response action (Table 4.2) suggested for the open ocean will be more important because of the increased potential for the bitumen to re-float in sticky clumps. Any re-floated bitumen will provide a threat to wildlife due to the greater number of birds and marine mammals present in the nearshore environment.

Furthermore, the potential for re-floated bitumen to come ashore as an oil slick or in the form of sticky tar patties and tar mats is much greater. Because shoreline cleanup is more costly and shoreline impacts increase the potential for natural resource and third-party damages, there should be an increased effort to respond to recover re-floated bitumen as quickly as possible.

The containment and recovery of re-floated "clumps" will follow the same strategy and, to a large degree, will use the same equipment available in the open-water marine environment. Response times should be faster for spills in nearshore waters, potentially increasing the effectiveness of on-water containment and recovery operations. If available, the viscous oil recovery equipment developed for Orimulsion may enhance the recovery rate for resurfaced Group V oil (Garcia Tavel et al., 1997). However, in order to make a difference, equipment such as the Tar Hawg, Oriboom, and other devices must be stockpiled in sufficient quantity at locations where the greatest potential for emulsified fuel spills exists (Middleton et al., 1995). Furthermore, local responders need to become familiar with the effective operation of this equipment. While existing equipment resources serving other oil transporters and storage facilities will satisfy the containment and recovery of resurfaced Group V oil requirements under the Oil Pollution Act of 1990 (OPA-90), especially if enhanced with the specially designed equipment for Orimulsion, the concern will be for the special on-water storage requirements necessary for these types of oils. It would seem that the viscosity of this product will mandate that all tank barges and tanks on skimming vessels be heated if off-loading of filled tanks is considered during a continuing on-water response operation.

Since there is greater potential for dispersed bitumen to re-float in the coastal zone, dispersants must be considered for this environmental scenario. Limited

TABLE 4.2 Effectiveness of Response Techniques for Spills of Emulsified Fuels in Marine Nearshore Environments

Response Technique	Effectiveness for Emulsified Fuels	Special Issues
Monitoring and tracking	Same as for open-water marine	SMART protocols should be used and augmented with large-volume water sampling systems to analyze dissolved and dispersed bitumen PAH concentrations
Dispersant application	Product disperses when spilled; dispersant effectiveness on re-floated bitumen will be very low	Application of dispersant in an attempt to redisperse re-floated bitumen may be ineffective. More study and field testing are required
Mechanical re-floating of dispersed emulsified oil	Because of faster response time for deployment of deep-skirted boom, this option may be more feasible than in open-ocean. However, low effectiveness levels would be expected	Increased tidal currents and wave energy in nearshore areas may reduce effectiveness
Long-skirted containment boom	Because of faster response time, this may be more effective in containing bitumen prior to extensive lateral dispersion	Recommended booming strategy for containing dispersed bitumen is a free-floating O-type configuration
Skimming	Faster response time for deep-skirted boom and skimmer deployment, increased potential for natural re-floating, increased on-station time for spotter aircraft	Same as for open-water marine
On-water storage	Same as for open-water marine	
Shoreline protection	Same basic strategy as for heavy oils. Expanded use of geotech materials, deep-skirted boom, and other products such as visqueen to cover shorelines prior to impact	Ability to carry out exclusion, deflection, and diversion of dispersed bitumen may require further study
Shoreline cleanup	Same basic strategy as for other heavy oils. Solvents or chemicals may enhance cleanup of some shoreline types	Preferred use of salt water in pressure washing to remove bitumen from substrates will increase wear and tear on equipment

TABLE 4.2 Continued

Response Technique	Effectiveness for Emulsified Fuels	Special Issues
Wildlife	Same strategy as for other oils	Studies required to determine the efficiency and toxicity of preferred cleaning agent
Dredging	Limited to submerged areas where sunken oil would accumulate	Many operational issues in regard to water depth, storage, disposal of recovered oily water and sediments, sea conditions, and currents
Diver-directed pumping and vacuum systems	Limited to submerged areas where sunken oil would accumulate	Visibility, water depth, storage and disposal of recovered oily water and sediments, sea conditions, and currents
Disposal	Same as for open-ocean	

studies have indicated that applying dispersant to re-floated bitumen will cause some redispersal (Oil Spill Response Limited, 1989). However, more work would have to be completed to determine if the dispersant application would be efficient and effective.

Shoreline protection and cleanup must also be considered in the coastal environment. For the most part, shoreline protection and cleanup of emulsified fuels will be conducted in the same manner as for other oil products (Owens and Sergy, 1999). There may also be a need to protect sensitive areas from shoreline impact if possible. Presumably this can be achieved by using long-skirted booms to exclude the oil from the zone to be protected, diverting it to a less sensitive shoreline for recovery or deflecting it back to open water (Morgan and Fernie, 1995; Owens and Sergy, 1999).

Once weathered bitumen is stranded on the shoreline, chemical agents enhance its removal (Guénette et al., 1998). However, further testing, particularly with agents accepted by the U.S. EPA for use in the United States, should be conducted to determine their efficiency in removing stranded bitumen coatings from various shoreline types and to study the fate and effects of the released bitumen.

As with other types of heavy oil, mechanized beach cleaners and other types of mechanical or manual recovery equipment are efficient in cleaning up stranded weathered bitumen from certain shorelines (Clement et al., 1997). Dredging or other types of underwater recovery must also be considered as a response option, because oil-sediment tar mats may be deposited in offshore depressions. Again, these operations will be similar to those used for spills of other types of heavy oil (National Research Council, 1999).

Because of the increased potential for contact with wildlife cleaning, of seabirds particularly, will become an issue. Even though a suitable agent has been proposed for cleaning wildlife (Gauvry and Miller, 1995), the effect of weathered Orimulsion on wildlife and the ability to clean contaminated wildlife for return to their habitat remain to be seen.

The report by Gauvry and Miller (1995) called for more study to determine the ability of the preferred cleaning agent to remove Orimulsion from birds and other species and to determine the toxicity of the cleaning agent itself to various types of wildlife.

ESTUARINE (BRACKISH WATER) RESPONSE

Only a small fraction of the spilled emulsified fuel is expected to re-float in an estuarine environment. The response strategies for estuaries will follow very closely those strategies discussed for the open-ocean, because the majority of the bitumen droplets will remain in suspension or sink (Table 4.3).

However, because many estuaries in the United States are also major port areas, the response time to spills should be much faster than for spills in the open-ocean or in coastal zones. Therefore, some of the response techniques discussed for the open-ocean and nearshore environments are appropriate for estuaries as well, and they may be more efficient because of the increased speed of response and access to shoreside support resources.

Estuaries do have zones of low flow where suspended particles may settle out, increasing the potential need for dredging or underwater vacuum recovery operations. However, the issue would be whether there is enough accumulation of "oil" to warrant bulk oil removal. Dredging and diver-assisted pumping or other types of underwater recovery operations can be very effective; however, they are also very expensive and require handling of large volumes of potentially contaminated water and sediment. Furthermore, dredging is not a commonly used spill response technique, and the required equipment may not be readily available.

NONTIDAL RIVER RESPONSE

Because of the fresh water and current conditions in a river, an emulsified fuel spill is expected to remain in suspension and to become more dispersed as it spreads further downstream. Therefore, tracking of the plume will require water column monitoring, either with grab water samples (limited spatial coverage, very slow) or field detectors such as fluorometers.

Even where trajectory models are available, they must be validated with field data on the location of the main body of the plume. Without preplanning that

TABLE 4.3 Effectiveness of Response Techniques for Spills of Emulsified Fuels in Estuaries (Brackish Water)

Response Technique	Effectiveness for Emulsified Fuels	Special Issues
Monitoring and tracking	With so little surface oil, it will be difficult to track the dispersed oil plume using overflights. Water sampling or remote sensing methods will be required, which have significant limitations	Gradients in water salinity over space and time, will have significant influence on ability to model the dispersed oil plume
Dispersant application	Not applicable because so little bitumen is expected to float or re-float	
Mechanical refloating and FAF of dispersed emulsified oil	Even though response times should improve, containing dispersed bitumen in tidal currents while attempting to re-float it will be difficult	If suspended plume can be diverted to a quiet backwater, this may be an option
Long-skirted containment boom	Because of tidal currents, even tear-dropped boom drifting with the current may not be able to contain dispersed droplets for recovery. May consider diverting dispersed bitumen plume to quiet water for settling-dredging or re-floating operations	Set and drift of vessels supporting re-floating and skimming operations will cause increased entrainment of bitumen from containment boom
Skimming	Because of improved response times and less area, skimming operations should be as effective for floating emulsified fuels as would be expected for other oils	
On-water storage	Same as open-ocean or nearshore	
Shoreline protection	Due to freshwater influence and less potential for "clumping" there may be less shoreline impact	
Shoreline cleanup	Due to freshwater influence and less potential for "clumping" there may be less shoreline impact	
Dredging	Same as nearshore	
Diver-directed pumping and vacuum systems	Same as nearshore	
Disposal	Same as open-ocean	

includes purchase, training, and maintenance of the necessary equipment, the monitoring of emulsified oil in rivers will be very difficult. There are at least 75 hazardous substances currently being transported in bulk on tankers and tank barges throughout the United States that present these same problems when spilled (U.S. Code of Federal Regulations). As discovered during a 1996 survey of resources available for response to releases of bulk hazardous substances transported by water, there may be a shortage of readily available equipment and personnel trained to monitor and sample products that either disperse, dissolve, or sink (Chemical Transportation Advisory Committee, 1996-1997). However, if and when the OPA-90 regulations for hazardous substances are finalized, they may include provisions that require transporters to have the ability to readily provide monitoring and sampling resources and expertise, and this situation will improve.

There will be very little opportunity for suspended droplets to re-float in this environment; therefore, there will be little or no opportunity to recover bitumen from the water surface using skimmers (see Table 4.4). The droplets will remain in suspension until the surfactant degrades, so they will be widely distributed before they attach to other suspended matter and sink. Therefore, it is doubtful that there will be sufficient accumulations in depression and behind natural and man-made structures to substantiate a dredging or diver-directed pumping or vacuum system recovery. The most feasible response in this environment will be to deploy exclusion or deflection long-skirted booms or silt fences at water intakes and other sensitive sites downstream.

Alternatively, downstream locations where the bitumen droplets might settle out (e.g. low-flow areas, backwater sloughs) and dredging-pumping-vacuum operations might be effective could be identified. If sufficient quantities of dispersed bitumen can be diverted to these quiet areas, then the PNP Re-floater or FAF principle could be considered as a response option. Based on limited experiments, the PNP Re-floater device is about 30 to 60 percent effective in re-floating dispersed bitumen in freshwater environments (Hvidbak and Masciangioli, 2000). Although recovery is lower than for salt water, laboratory testing indicates a good recovery of dispersed re-floated bitumen in fresh water. However, aeration is essential since recovery only occurs at the surface. If the sorbent or polymer material is delivered as part of the aeration process, efficiency may be increased even further (M3 Inc., 1996).

FRESHWATER QUIESCENT RESPONSE

A discharge of emulsified fuel in a freshwater, quiescent setting, such as a port facility, will offer the best opportunity for containing and recovering the spilled bitumen or oil. A facility that receives routine cargo of emulsified fuel can plan for spill events and provide the necessary equipment needed for containing and recovering it.

TABLE 4.4 Effectiveness of Response Techniques for Spills of Emulsified Fuels in Nontidal Rivers

Response Technique	Effectiveness for Emulsified Fuels	Special Issues
Monitoring and tracking	Visual techniques will not be effective in most rivers. Most effective would be fluorometers deployed on boats or at one location (e.g., water intake)	Real-time river flow data are available to predict spread of the dispersed plume. Would require preplanning for use of water column monitors. May be shortage of readily available sampling equipment and qualified technicians
Dispersant application	Since emulsified fuels will remain in suspension in fresh water, dispersant will not be considered	None
Mechanical re-floating of dispersed emulsified oil	In most cases, not a viable or efficient option. However, if dispersed bitumen can be diverted to a quiet water area, it may be tried	Effectiveness of PNP Re-floater and FAF has to be validated for fresh water
Long-skirted booming	May consider diversion or exclusion booming to protect water intakes or sensitive areas. Can probably be effective only in slow-current environments	Specially designed booms with semi-permeable skirts might be more effective than regular booms
Skimming	Use only in conjunction with PNP and FAF operations	None
Sorbents and polymers	May be effective, if used with aeration	Containment will be an issue using these techniques
On-water storage	Required for skimming conducted in conjunction with PNP and FAF operations	None
Shoreline protection	Not applicable because no shoreline stranding should occur	None
Shoreline cleanup	Not applicable because no shoreline stranding should occur	None
Dredging	Low effectiveness because of low potential for accumulation of recoverable amounts of bottom oil	River currents should keep particles in suspension and enhance spreading
Diver-directed pumping and vacuum systems	Low effectiveness because of low potential for accumulation of recoverable amounts of bottom oil	Low visibility in rivers may make diver operations less effective
Disposal	Same as open ocean	None

Furthermore, although the equipment designed for Group IV and V oils may not be as efficient as specially designed equipment, it will be more available and perhaps more rapidly deployed in this environment. Because many transfer facilities are situated in at least semi-protected areas, it is likely that the berth will be pre-boomed (or can immediately be boomed) with a deep-skirted boom or silt curtain following a spill. In this case, the majority of the bitumen droplets will remain in suspension until they sink to the bottom of the berth or are re-floated. Although the PNP Re-Floater may achieve only 30 percent success in fresh water, a fully contained spill would provide more time to work the dispersed bitumen, and a greater percentage may be obtained (see Table 4.5).

A fully contained spill at a shoreside facility would also allow greater opportunity for the use of FAF system technology. It would allow the dispersed plume to be pumped to shoreside tanks where FAF would allow separation, floatation, and recovery.

This scenario also provides the best opportunity for carrying out dredging or diver-assisted pumping operations to recover the bitumen droplets that settle on the bottom. Once on the bottom the bitumen will likely remain until recovery can take place. The shallow water depths in a port area will also be conducive to dredging operations.

ON LAND NEAR WATER RESPONSE

When spilled on land, fresh Orimulsion will behave like a viscous liquid, with the potential for penetration into porous substrates. Response and cleanup methods would be similar to those for conventional heavy oils. As it weathers and becomes sticky, the degree of penetration will decrease, and cleanup will likely involve complete removal of the surface oil and oiled sediments. Even though the risk of groundwater contamination is low, groundwater cleanup methods would be needed for the water-soluble fraction contained in the water component.

In this type of spill, the emulsified fuel will remain in suspension when in water. Containment booming using regular containment boom and filter-fence-type material should be possible (see Table 4.6), because of the reduced current and very shallow-water conditions expected in a wetland habitat. Dikes and berms could also be constructed to isolate the spill area from other waterways and reduce migration. Depending on the amount of product spilled, the size of the area impacted, and the sensitivity of the wetland, the area could be isolated and pumped dry for manual or heavy equipment recovery of oil or bitumen from the substrate. The spill site could then be treated by burning or land farming. To ensure that only clean water was discharged downstream the effluent from any pumping operation to a dry out area could be routed through a filter box arrangement, or FAF or a PNP Refloater pumping system could be used to pump out the marsh water. In areas where the land is alternately wet and dry, emulsified fuels,

TABLE 4.5 Effectiveness of Response Techniques for Spills of Emulsified Fuels in Freshwater Quiescent Environments

Response Technique	Effectiveness for Emulsified Fuels	Special Issues
Monitoring and tracking	Can be relatively effective because the point of exit can be the focus of water column monitoring	None
Dispersant application	Not applicable because none of the oil will float	None
Mechanical re-floating of dispersed emulsified oil	Tests indicate that only 30 percent effectiveness can be achieved in fresh water. However, logistics available in this scenario may make it a viable option	None
FAF system	Shoreside facilities and ability to contain dispersed bitumen plume make this a viable option for this scenario	Oil water treatment facility and/or decanting authority would be necessary
Long-skirted containment boom	Predeployed deep-skirted boom should be considered for transfer facilities. For emulsified fuel that escapes the encapsulated area, river environment techniques would apply	Facility response planning allows for advanced containment strategy and equipment staging
Sorbents and polymers	Same as rivers	Containment should be improved in this environment
Storage	Should not be limiting because of many shoreside options	Facility response planning allows for advanced containment strategy and equipment staging
Shoreline protection	Limited to areas where re-floating operations take place	None
Shoreline cleanup	Limited to areas where re-floating operations take place	None
Dredging	Oil or bitumen that sinks should be contained in the immediate vicinity of berth and readily recovered	None
Diver-directed pumping and vacuum systems	Oil or bitumen that sinks should be contained in the immediate vicinity of berth and readily recovered	Bottom topography can be highly variable, so divers may have to direct recovery to zones of higher accumulation

TABLE 4.6 Effectiveness of Response Techniques for Spills of Emulsified Fuels on Land Near Water

Response Technique	Effectiveness for Emulsified Fuels	Special Issues
Monitoring and tracking	Can be highly effective since plume is highly visible	None
Dispersant application	Not applicable because none of the oil will float	None
Mechanical refloating of dispersed emulsified oil	Dependent on water depth. May be considered part of any pumping operation intended to increase flow from marsh or to dry area for recovery of product that sinks	None
Containment boom	Regular boom, filter fence, and deep-skirted boom should be considered depending on water depths	Access will be an important consideration
Berms or dikes	Install berm or dike to contain spill within marsh area already impacted	Environmental concern regarding impact of stopping flow of water through marsh
Skimming	Not effective because none of the bitumen is expected to float	None
Sorbents and polymers	Could be moderately effective where plume can be concentrated to flow through a restricted area	None
Storage	May be difficult in areas of limited access	None
Shoreline protection	Minimal shoreline protection strategy required	None
Excavation or dredging	Method dependent on water depth and soil conditions. Removal of contaminated sediments using heavy equipment or manual labor	Disposal of large volumes of material
Shoreline cleanup	Shoreline cleanup will be required on any surface where the product flowed across soil or vegetation and dried. Normal oil shoreline cleanup techniques will apply	Access will define type of cleanup technique to large extent

TABLE 4.6 Continued

Response Technique	Effectiveness for Emulsified Fuels	Special Issues
Dredging	Same as for estuarine environment. Amount spilled, area impacted, type of marsh environment, and water depth will be critical	Access will be critical for defining type of dredging equipment that may be used
Diver-directed pumping and vacuum system	Same as dredging above	None
Excavation	Pump impacted marsh or wetland area dry, and excavate oil or bitumen that has sunk and accumulated	Environmental effect of pumping marsh dry
In situ burning	Pump impacted marsh or wetland area dry, and burn oil or bitumen that has sunk and accumulated	Burning agent will have to be used to support in situ burn of bitumen. Environmental effect of pumping marsh dry
Land farming or bioremediation	Pump impacted marsh or wetland area dry, and land-farm or fertilize to enhance biodegradation	Environmental effect of pumping marsh dry

and bitumen in particular, are less likely to leach into the substrate than other heavy oils (Wood, 1996).

SUMMARY

Because emulsified fuels are essentially predispersed, the most likely response actions will be monitoring of the dispersed plume and recovery of any re-floated bitumen. Recovery efforts are likely to be even lower than for traditional spills because so little of the oil is expected to re-float. Therefore, as for all types of spills, the most effective strategy, is prevention.

Many of the proposed strategies for responding to spills of emulsified fuels are likely to have low initial effectiveness for the following reasons:

- They have not been tested under realistic conditions.
- The equipment may not be readily available at all necessary locations.
- Responders are not familiar with the equipment and strategies available for emulsified fuel spills

- There have not been spills where the proposed strategies could be tested and made more effective under field conditions.
- For example, mechanical re-floating of dispersed bitumen has been demonstrated in small intentional spill scenarios and is suggested as a practical response to a spill of emulsified fuel.

However, only actual experience will determine if it is practical to re-float any type of dispersed oil and, if required, to determine whether the methodologies being suggested are logistical and practical for likely spill scenarios. Another proposed strategy is diversion or deflection of a dispersed plume using deep-skirted booms; field tests should be conducted to validate this strategy and determine the physical and environmental limitations. Tests with trawl-type nets for containing and recovering floating weathered bitumen indicate that smaller-mesh nets might make the systems more effective. Further tests are needed to determine if towing smaller-mesh nets is practical. **Realistic field tests[1] and refinement of the more innovative methods for containment and recovery of emulsified fuels are required before these methods can be part of realistic response plans.**

---

[1]Environmental restrictions on the release of possibly toxic substances may contstrain certain types of field tests.

References

AEA Technology. 1996. Investigations into Landspills of Orimulsion. Report no. AEA/WMES/RMAF/20206001. Oxfordshire, UK.[1]

Ahel, M., E. Moinar, S. Ibric, and W. Giger. 2000. Estrogenic metabolites of alkylphenol polyethoxylates in secondary sewage effluents and rivers. Water Science and Technology 42(7-8):15-22.

American Petroleum Institute. 1999. Beyond 2000, balancing perspectives. In: Proceedings of the 1999 International Oil Spill Conference, Seattle, Washington. Publication no. 4686B. Washington, D.C: American Petroleum Institute.

Anderson, J. W., J. M. Neff, B. A. Cox, H. E. Tatem, and G. M. Hightower. 1974a. Characteristics of dispersions and water-soluble extracts of crude and refined oils and their toxicity to estuarine crustaceans and fish. Marine Biology 27:75-88.

Anderson, J. W., J. M. Neff, B. A. Cox, H. E. Tatem, and G. M. Hightower. 1974b. The effects of oil on estuarine animals: Toxicity, uptake and depuration, respiration. In: Pollution and Physiology of Marine Organisms, F. J. and W. B. Vernberg, (eds.). New York: Academic Press, Inc., pp. 285-310.

Anderson, J. W., S. L. Kiesser, R. M. Bean, R. G. Riley, and B. L. Thomas. 1981. Toxicity of chemically dispersed oil to shrimp exposed to constant and decreasing concentrations in a flowing system. In: Proceedings of the 1981 Oil Spill Conference. Washington, D.C.: American Petroleum Institute, pp. 69-75.

Anderson, J. W., S. L. Kiesser, D. L. McQuerry, R. G. Riley, and M. L. Fleischmann. 1984. Toxicity testing with constant or diluting concentrations of chemically dispersed oil. In: E. A. Thomas (ed.). Oil Spill Chemical Dispersant: Research, Experience, and Recommendations. STP 840. Philadelphia: American Society for Testing and Materials, pp. 14-22.

---

[1]Orimulsion® is a registered trademark belonging to Bitúmenes Orinoco, S.A (PDVSA-Bitor) and licensed to Bitor America Corporation.

REFERENCES

Anderson, J. W., R. Riley, S. Kiesser, and J. Gurtisen. 1987. Toxicity of dispersed and undispersed Prudhoe Bay crude oil fractions to shrimp and fish. In: Proceedings of the 1987 Oil Spill Conference. Washington, D.C.: American Petroleum Institute, pp. 235-240.

Ankley, G. T., R. J. Erickson, G. L. Phipps, V. R. Mattson, P. A. Kosian, B. R. Sheedy, and J. S. Cox. 1995. Effects of light-intensity on the phototoxicity of fluoranthene to a benthic macroinvertebrate. Environmental Science and Technology 29 (11):2828-2833.

Armsworthy, S. L., P. J. Canford, G. H. Tremblay, and K. Lee. 1999. Effects of Orimulsion on Food Acquisition and Growth of Sea Scallops. Nova Scotia, Canada: Bedford Institute of Oceanography, Marine Environmental Science Division, Department of Fisheries and Oceans.

Ault, J. S., M. A. Harwell, and V. Myers, (eds.) 1995. Comparative Ecological Risk Assessment. Technical Support Document for the Comparison of the Ecological Risks to the Tampa Bay Ecosystem from Spills of Fuel Oil #6 and Orimulsion. Final Report, Volume II. Miami, Florida: University of Miami, Center for Marine and Environmental Analyses, Rosenstiel School of Marine and Atmospheric Science.

Bitor America Corporation. 1997. Orimulsion Containment and Recovery Test Carried Out in Puerto La Cruz, Anzoategui State, Venezuela. Final Report, Version 1.1., Boca Raton, Florida: Bitor America Corporation.*

Bitor America Corporation. 1999. Orimulsion Spill Response Manual. Volume II: Protection and Cleanup of Marine Shorelines. Boca Raton, Florida: Bitor America Corporation.*

Bitúmenes Orinoco, S.A (PDVSA-Bitor) a. Physical and Chemical Properties of Orimulsion-400. Boca Raton, Florida: Bitor America Corporation.*

Bitúmenes Orinoco, S.A (PDVSA-Bitor) b. Orimulsion—Behavior of Orimulsion-400 in Water. Boca Raton, Florida: Bitor America Corporation.*

Bjornestad, E., A. R. Pedersen, and S. Bowadt. 1998. Ecotoxicological characterization of Orimulsion 400: Summary Report. VKI Project Number 11020. Hørsholm, Denmark: VKI.

Boesch, D. F., and N. Rabalais, (eds.). 1987. Long-term Environmental Effects of Offshore Oil and Gas Development. New York: Elsevier Applied Science.

Bowadt, S., B. Jensen, and K. Lund. 1998. Development of an Electrospray Determination of Alcohol Ethoxylates in Water Samples down to Trace Levels. Hørsholm, Denmark: VKI.

Brown, J., H. Fuentes, R. Jaffé, and V. Tsihrintzis. 1995. Comparative evaluation of physical and chemical fate processes of Orimulsion and fuel oil No. 6 in the Tampa Bay marine environment. In: Comparative Ecological Risk Assessment-Technical Support Document for the Comparison of the Ecological Risks to the Tampa Bay Ecosystem from Spills of Fuel Oil #6 and Orimulsion-Final Report, Ault, J. S., M. A. Harwell, and V. Myers, (eds.). Miami, Florida: University of Miami, Center for Marine and Environmental Analyses, Rosenstiel School of Marine and Atmospheric Science.

Calabrese, E., P. Kostecki, and L. Baldwin. 1995. Toxicological assessment of Fuel Oil Number 6 and Orimulsion: recommendations for future testing. In: Comparative Ecological Risk Assessment-Technical Support Document for the Comparison of the Ecological Risks to the Tampa Bay Ecosystem from Spills of Fuel Oil #6 and Orimulsion—Final Report, Volume II, Ault, J. S., M. A. Harwell, and V. Myers, (eds.). Miami, Florida: University of Miami, Center for Marine and Environmental Analyses, Rosenstiel School of Marine and Atmospheric Science.

Chemical Transportation Advisory Committee. 1996-1997. Subcommittee for Hazardous Substance Response Planning. Findings of the working group for evaluating existing response capability.

Clement, F., P. Gunter, and D. Oland. 1997. Trials of recovery and clean-up techniques on bitumen derived from Orimulsion. In: Proceedings of the 1997 International Oil Spill Conference, Fort Lauderdale, Florida. Washington, D.C.: American Petroleum Institute, pp. 89-93.

Conover, R. J. 1971. Some relations between zooplankton and bunker C oil in Chedabucto Bay following the wreck of the tanker *Arrow*. Journal of the Fisheries Research Board of Canada 28(9):1327-1330.

Corner, E. D. S., R. P. Harris, C. C. Kilvington, and S. C. M. O'Hara. 1976a. Petroleum compounds in marine food web: short-term experiments on the fate of naphthalene in *Calanus*. Journal of the Marine Biological Association of the United Kingdom 56(1):121-133.

Corner, E. D. S., R. P. Harris, K. J. Whittle, and P. R. Mackie. 1976b. Hydrocarbons in marine zooplankton and fish. In: Effects of Pollutants on Aquatic Organisms, A. P. M. Lockwood (ed.). London: Cambridge University Press, pp. 71-105.

Crosbie, A., and A. Lewis. 1998a. Laboratory Studies on the Spill Behavior of Orimulsion-100 and Orimulsion-400. Report No. AEAT-3381. Oxfordshire, UK: National Environmental Technology Centre.

Crosbie, A., and A. Lewis.1998b. Studies on the Spill Behavior of Orimulsion-100 and Orimulsion-400 Using Standard WSL and IFP Tests. Report No. AEAT-3481. Oxfordshire, U.K.: National Environmental Technology Centre.

Davis, J.W., and C. L. Carpenter. 1997. Environmental assessment of the alkanolamines. In: Reviews of Environmental Contamination and Toxicology 49. New York: Springer Verlag.

Di Corcia, A., C. Crescenzi, A. Marcomini, and R. Samperi. 1998. Liquid chromatography electrospray mass spectrometry as a valuable tool for characterizing biodegradation intermediates of branched alcohol ethoxylate surfactants. Environmental Science & Technology 32(5):711-718.

Deis, D. R., N. Garcia Tavel, C. Villoria, G. Febres Ortega, P. Masciangioli, M. A. Jones, and G. R. Lee. 1997. Orimulsion: Research and testing and open water containment and recovery trials. In: Proceedings of the 1997 International Oil Spill Conference, Fort Lauderdale, Florida. Washington D.C.: American Petroleum Institute, pp. 459-467.

Dorn, P. B., J. P. Salanitro, S. H. Evens, and L. Kravetz. 1993. Assessing the aquatic hazard of some branch and linear nonionic surfactants by biodegradation and toxicity. Environmental Toxicology and Chemistry 12(10):1751-1762.

Dorn, P. B., J. H. Rodgers Jr., W. B. Gillespie Jr., R. E. Lizotte Jr., and A. W. Dunn. 1997. The effects of a C_{12-13} linear alcohol ethoxylate surfactant on periphyton, macrophytes, invertebrates and fish in stream mesocosms. Environmental Toxicology and Chemistry 16(8):1634-1645.

Duffy, L. K., R. T. Bowyer, J. W. Testa, and J. B. Faro. 1994. Chronic effects of the Exxon Valdez oil spill on blood and enzyme chemistry of river otters. Environmental Toxicology and Chemistry 13(4):643-647.

Energy Information Administration (EIA). 2000. Annual electric generator report. Annual Energy Review 1999. DOE/EIA-0384(99).

Energy Information Administration. 2001. Annual energy outlook 2001: With projections to 2020.

Environment Canada. 2001. PAH Concentrations in Crude Oils and Refined Products. Unpub. Environment Canada, Ottawa.

European Commission. In press. Draft European union risk assessment report: Nonylphenol and phenol, 4-nonyl-, branched.

Febres, G. A., J. A. Gonçalvez, and J. Vilas. 1995. Fate and Behavior of Orimulsion Spilt in Sea Water. Intevep, S.A.

French, D. P., and D. Mendelsohn. 1995. Development and application of an Orimulsion spill fate and effects model. In: Proceedings of the 18th Arctic and Marine Oilspill Program, Technical Seminar, June 14-16, 1995, Edmonton, Alberta, Canada. Environment Canada, Volume 2, pp. 769-792.

French, D., M. Reed, K. Jayko, S. Feng, H. M. Rines, S. Pavignano, T. Isaji, S. Puckett, A. Keller, F. W. French III, D. Gifford, J. McCue, G. Brown, E. MacDonald, J. Quirk, S. Natzke, R. Bishop, M. Welsh, M. Phillips, and B. S. Ingram. 1996. The CERCLA type A natural resource damage assessment model for coastal and marine environments (NRDAM/CME), technical documentation, Vol. I-V, final report, Contract number 14-0001-91-C-11. Office of Environmental Policy and Compliance, U.S. Department of the Interior: Washington, D.C.

French, D. P., H. Rines, and P. Masciangioli. 1997. Validation of an Orimulsion spill fates model using observations from field test spills. In: Proceedings of the 20th Arctic and Marine Oilspill Program Technical Seminar. Environment Canada, pp. 933-961.

French, D. 2000. Final Report: Update of the SIMAP Model for Orimulaion-400. Bitor Europe Limited, pp 66.*

French McCay, D., and C. Galagan. 2001. Modeling of Orimulsion and Heavy Fuel Oil Spills in Delaware River and Bay. Report No. 00-050. Narragansett, Rhode Island: Applied Science Associates.

Fry, D. M., and L. J. Lowenstine. 1985. Pathology of common murres and Cassin's auklets exposed to oil. Archives of Environmental Contamination and Toxicology 14:725-737.

Garcia Tavel, N., C. Villoria, G. Febres Ortega, P. Mansciangioli, M. Jones, and G. Lee, 1997. Orimulsion research and testing and open water containment and recovery trials. In: Proceedings of the 1997 International Oil Spill Conference, Fort Lauderdale, Florida, April 7-10. Washington, D.C.: American Petroleum Institute, pp. 459-467.

Gauvry, G. A., and E. A. Miller. 1995. Florida Power and Light Research Program on Orimulsion Contamination of Avian Species and Feasibility of Cleaning. Tri-State Bird Rescue and Research, Inc.*

Golder and Associates, Inc., 2001. Golder—Surfactant Fact Sheet: Orimulsion-400 Surfactant Package, 013 7572, September 26, 2001. Boca Raton, Florida: Bitor America Corporation.*

Golder Associates Geoanalysis S.r.l. 1999. Orimulsion-400 and Fuel Oil No.6: A Comparative Study of Aquatic Ecotoxicology. Report No. 992210/3683. Rome, Italy: Bitor Italia S.r.l.*

Guénette, C., G. Sergy, and B. Fieldhouse. 1998. Removal of stranded bitumen from intertidal sediments using chemical agents. Phase I: Screening of chemical agents. Report no. EE-162. Ottawa, Canada: Environment Canada.

Harper, J. R., and M. Kory. 1997. Orimulsion Shoreline Studies Program Sediment Interaction Experiments. Manuscript Report Environmental Protection Directorate, Ottawa, Ontario.

Harwell, M. A., and I. C. Johnson. 2000. Appendix H: Comparative ecological risk assessment of Orimulsion spills in Tampa Bay: Supplemental information. Golder Associates, Inc., Gainesville, Florida.

Hemond, H. F., and E. J. Fechner. 1994. Chemical Fate and Transport in the Environment. San Diego: Academic Press, Inc.

Hvidbak, F., and P. Masciangioli. 2000. The development and test of equipment for rapid refloatation of spilled Orimulsion. In: Proceedings of the 23rd Arctic and Marine Oilspill Program Technical Seminar, Environment Canada 1:469-480.

Johnson, I., P. Jokuty, K. Doe, and S. Blenkinsopp. 1998. A review of methods used in the preparation and analysis of Orimulsion related samples (Part 2-Aquatic toxicology). In: Proceedings of the 23rd Arctic and Marine Oilspill Program (AMOP), Technical Seminar. Ottawa, Canada: Environment Canada, pp. 261-279.

Johnson, I. C. 1998. Summary of the environmental fate and effects of Orimulsion-400. Boca Raton, Florida: Bitor America Corporation.*

Jokuty, P., S. Whiticar, M. Fingas, Z. Wang, K. Doe, D. Kyle, P. Lambert, and B. Fieldhouse. 1995. Orimulsion: Physical properties, chemical composition, dispersibility and toxicity. Report no. EE-154. Ottawa, Ontario: Environment Canada.

Jokuty, P., S. Whiticar, Z. Wang, K. Doe, B. Fieldhouse, and M. Fingas. 1999. Orimulsion-400: A Comparative Study, Manuscript Report no. EE-160, Ottawa, Ontario: Environmental Protection Service, Environment Canada.

Karsa, D. R. (ed.). 1998. New Products and Applications in Surfactant Technology. Sheffield, U.K.: Academic Press Ltd.

Kravetz, L., J. P. Salanitro, P. B. Dorn, and K. F. Guin. 1991. Influence of hydrophobe type and extent of branching on environmental response factors of nonionic surfactants. Journal of the American Oil Chemists Society 68:610-618.

REFERENCES

Lapham, L., L. Proctor, and J. Chanton. 1999. Using respiration rates and stable carbon isotopes to monitor the biodegradation of Orimulsion by marine benthic bacteria. Environmental Science & Technology 33(12):2035-2039.

Lee, R. 1975. Fate of petroleum hydrocarbons in marine zooplankton. In: Proceedings of the 1975 Conference on Prevention and Control of Oil Pollution. Washington, D.C.: American Petroleum Institute, pp. 549-553.

Leighton, F. A., D. B. Peakall, and R. G. Butler. 1983. Heinz-body hemolytic-anemia from ingestion of crude oil: A primary toxic effect in marine birds. Science 220:871-873.

Li, M., and C. Garrett. 1998. The relationship between oil droplet size and upper ocean turbulence. Marine Pollution Bulletin 36(12):961-970.

Lunel, T. 1993. Dispersion: Oil droplet size measurements at sea. In: Proceedings of the 16th Arctic and Marine Oil Spill Program (AMOP) Technical Seminar. Ottawa, Canada: Environment Canada, pp. 1023-1056.

M3 Incorporated. 1996. Orimulsion: Applicability of Sorbents and Polymers for Spill Recovery in Freshwater Environments. Orimulsion Sorbent Evaluation Phase I, Vol. FW002.*

Marcomini, A., M. Zanette, G. Pojana, and M. J. F. Suter. 2000a. Behavior of aliphatic alcohol polyethoxylates and their metabolites under standardized aerobic biodegradation conditions. Environmental Toxicology and Chemistry 19(3):549-554.

Marcomini, A., G. Pojana, C. Carrer, L. Cavalli, G. Cassani, and M. Lazzarin. 2000b. Aerobic biodegradation of monobranched aliphatic alcohol polyethoxylates. Environmental Toxicology and Chemistry 19(3):555-560.

Markarian, R. K., K. W. Pontasch, D. R. Peterson, and A. I. Hughes. 1989. Comparative toxicities of selected surfactants to aquatic organisms. In: Review and Analysis of Environmental Data on Exxon Surfactants and Related Compounds, Technical Report. East Milestone, N.J.: Exxon Biomedical Sciences, Inc.

Meador, J. P., J. E. Stein, W. L. Reichert, and U. Varanasi. 1995. Bioaccumulation of polycyclic aromatic hydrocarbons by marine organisms. Reviews of Environmental Contamination and Toxicology 143:79-165.

Michel, J., D. Scholz, C. B. Henry, and B. L. Benggio. 1995. Group V fuel oils: Source, behavior, and response issues. In: Proceedings from the 1995 International Oil Spill Conference, Long Beach, California. Washington, D.C.: American Petroleum Institute.

Middleton, C., R. Brown, and J. Lucarelli. 1995. Report on Orimulsion Pollution Control Testing, Cape Canaveral Marine Services, Inc., Cape Canaveral, Florida.*

Miller, C. A., and R. K. Srivastava. 2000. The combustion of Orimulsion and its generation of air pollutants. Progress in Energy and Combustion Science 26(2):131-160.

Miller, C. A., K. Dreher, R. Wentsell, and R. J. Nadeau. 2001. Environmental Impacts of the Use of Orimulsion® Technology Assessment Program. Volume 1. Excecutive Summary, Basic Report, Appendix Q. EPA-600/R-01-056a.

Morgan, D. C., and W. A. Fernie. 1995. Report on the Orimulsion Pollution Control Testing Carried Out. OSSC. Southhampton, UK: Frank Ayles & Associates Limited.

National Energy Policy Development Group (NEPDG). 2001. Reliable, Affordable, and Environmentally Sound Energy for America's Future. Washington, D.C.: U.S. Government Printing Office. Available at: http://www.energy.gov/HQPress/releases01/maypr/energy_policy.htm.

National Oceanic and Atmospheric Administration. ADIOS Model Database.

National Research Council (NRC). 1985. Oil in the Sea: Inputs, Fates, and Effects. Washington, D.C.: National Academy Press.

National Research Council (NRC). 1999. Spills of Nonfloating Oils Risk and Response. Washington, D.C.: National Academy Press.

Naylor, C. G., J. P. Mieure, W. J. Adams, J. A. Weeks, F. J. Castaldi, L. D. Ogle, and R. R. Romano. 1992. Alkylphenol ethoxylates in the environment. Journal of the American Oil Chemists Society 69:695-703.

Neff, J. M., and J. W. Anderson. 1982. Response of Marine Animals to Petroleum and Specific Petroleum Hydrocarbons. London: Applied Science Publishers Ltd.

Oil Spill Response Limited. 1989. A Report on a Study to Determine Treatment Options Following Spillage of Orimulsion into the Marine and Fresh Water Environments. Southampton, U.K.: Oil Spill Service Centre.

Okubo, A. 1971. Oceanic diffusion diagrams. Deep Sea Research 18:789-802.

Ostazeski, S. A., S. A. Stout, and A. D. Uhler. 1998a. Testing and Characterization of Orimulsion 400. Volume 1. Technical Report. Final Report. Duxbury, Massachusetts: Battelle.*

Ostazeski, S. A., S. A. Stout, and A. D. Uhler. 1998b. Testing and Characterization of Orimulsion 400 Volume 2—Appendices. Duxbury, Massachusetts: Battelle.*

Owens, E. H., and G. A. Sergy. 1999. Orimulsion Spill Field Guide for the Protection and Cleanup of Marine Shorelines. Edmonton, Alberta: Emergency Science Division, Environment Canada, 194 pp.

Page, G. W., and H. R. Carter. 1986. Impacts of the 1986 San Joaquin Valley crude oil spill on marine birds in central California. Report No. 353, Point Reyes Bird Observatory.

Payne, J. R., T. J. Reilly, and D. P. French. 1999. Fabrication of a portable large-volume water sampling system to support oil spill NRDA efforts. In: Proceedings of the 1999 Oil Spill Conference. Washington, D.C.: American Petroleum Institute, pp. 1179-1184.

PDVSA–Intevep. (1998). Density of Orimulsion-400, ITS Bitumen, and Different Salinity Waters. Boca Raton, Florida: Bitor America Corporation.*

Potter, T., J. Wu, K. Simmons, P. Kostecki, and E. Calabrese. 1997. Chemical Characteristics of the Water Soluble Fraction of Orimulsion-in-Water Dispersions. Amherst, Massachusetts: University of Massachusetts, Department of Food Science and School of Public Health.

Proctor, L. M., E. Toy, L. Lapham, J. Cherrier, and J. P. Chanton. 2001. Enhancement of Orimulsion biodegradation through the addition of natural marine carbon substrates. Environmental Science & Technology 35(7):1420-1424.

Roland, S., P. Donkin, E. Smith, and E. Wraige. 2001. Aromatic hydrocarbon "humps" in the marine environment: Unrecognized toxins? Environmental Science and Technology 35:2640-2644.

Sommerville, M. 1999. Orimulsion containment and recovery. Pure and Applied Chemistry 71(1):193-201.

Spies, R. B., S. D. Rice, D. A. Wolfe, and W. A. Wright. 1996. Effects of the *Valdez* Oil Spill on the Alaska Coastal Environment. In: S. D. Rice, R. B. Spies, D. A. Wolfe, and W. A. Wright (eds.). Proceedings of the *Exxon Valdez* Oil Spill Symposium, American Fisheries Society Symposium No.18. American Fisheries Society: Bethesda, Maryland, pp. 1-16.

Stanford Research Institute. 1998. Chemical Economics Handbook. Menlo Park, California: Stanford Research Institute.

Stout, S. A. 1999. Predicting the Behavior of Orimulsion Spilled on Water. U.S. Department of Transportation Project No. 4120.11/UDI283. Groton, Connecticut: U.S. Coast Guard.

Swisher, R. D. 1987. Surfactant Biodegradation, 2nd ed. Surfactant Series, Vol. 18. New York: Marcel Dekker.

Talmage, S. S. 1994. Environmental and Human Safety of Major Surfactants Alcohol Ethoxylates and Alkylphenol Ethoxylates. Boca Raton, Florida: Lewis Publishers.

United States Coast Guard. 1996. Vessel Response Plans. Federal Register 61(9):1051-1107.

U.S. Code of Federal Regulations. Chapter 46, Parts 151, 153, and 154.

van de Plassche, E. J., J. H. M. de Bruijn, R. R. Stephenson, S. J. Marshall, T. C. J. Feijtel, and S. E. Belanger. 1999. Predicted no-effect concentrations and risk characterization of four surfactants: Linear alkyl benzene sulfonate, alcohol ethoxylates, alcohol ethoxylated sulfates, and soap. Environmental Toxicology and Chemistry 18(11):2653-2663.

VKI. 1997a. Biodegradability of a Mixture of GENAPOL X 159- and 75090 Monoethanolamine (7000/1580) Marine Closed Bottle Test. Hørsholm, Denmark: VKI.

VKI. 1997b. Biodegradability of 75090 Monoethalomine - Marine Closed Bottle Test (Test Period October 10, 1997 - November 11, 1997), GLP Study # 81047-2/064. Hørsholm, Denmark: VKI.

VKI. 1999. Orimulsion-400 and Heavy Fuel Oil: Assessment of Risks for Marine Accidents and the Environment During Transport to Asnæs Power Plant. Project No. 11896.

Wang, Z., and M. Fingas. 1996. Separation and characterization of Petroleum Hydrocarbons and Surfactant in Orimulsion Dispersion Samples. Environmental Science and Technology 30:3351-3361.

Wells, P., J. N. Butler, and J. Staveley Hughes, (eds.). 1995. Exxon Valdez Oil Spill: Fate and Effects in Alaskan Waters. Philadelphia, Pennsylvania: American Society for Testing Materials.

Witham, R. 1983. A review of some petroleum impacts on sea turtles. In: Proceedings of the Workshop on Cetaceans and Sea Turtles in the Gulf of Mexico: Study Planning for Effects of Outer Continental Shelf Development. Report no. USFWS/OBS. C. E. Kellis and J. K. Adams (eds.).

Wolfe, M., J. White, G. Gauvry, and S. Patton. 2001. Orimulsion/No. 6 fuel oil toxicity in mallards: Phase I. Gainesville, Florida: Golder Associates Inc.*

Wood, P. 1996. Investigations into Land Spills of Orimulsion. Oxfordshire, U.K.: AEA Technology, pp. 1-18.

---

* Contact Bitor America Corporation at the address below to obtain technical reports.

Bitor America Corporation
5100 Town Center Circle, Suite 301
Boca Raton, Florida 33486
(561) 392-0026
(561) 392-0490
http://www.bitoramerica.com/

Appendix A

Committee Biographies

Jacqueline Michel *(Chair)*
Research Planning, Inc.

Jacqueline Michel received her Ph.D. from the University of South Carolina in Geochemistry in 1980. Currently, Dr. Michel is the President of Research Planning, Inc. Dr. Michel is an expert in oil and chemical response and contingency planning. She has been the program manager providing specific support to NOAA's Hazardous Materials Response and Assessment Division since 1978. Dr. Michel is a member of the Committee on Oil in the Sea: Inputs, Fates, and Effects and was a member on the Committee on Marine Transportation of Heavy Oil.

Jack Anderson
Columbia Analytical Services

Jack Anderson received his Ph.D. in Biology from the University of California, Irvine in 1969. Currently, Dr. Anderson is a Principal Scientist in the Technology Department at Columbia Analytical Services, Inc. Dr. Anderson has over 29 years of experience investigating the fate and effects of petroleum hydrocarbons and other pollutants on the marine environment and marine organisms. Dr. Anderson has published numerous papers, book chapters and books on the effects of contaminants on marine organisms, and has also developed a new method of testing toxic and carcinogenic compounds in samples of water, tissue, soil and sediments.

Charles F. Bryan
U.S. Geological Survey and Louisiana State University

Fred Bryan received his Ph.D. in Zoology from the University of Louisville. Currently, Dr. Bryan is a Leader in the Louisiana Cooperative Fish and Wildlife Unit of the U.S. Geological Survey/Biological Resources Division at Louisiana State University. Dr. Bryan specializes in synoptic surveys on water quality, plankton, benthic invertebrates, and larval and adult fishes in the lower Mississippi and Atchafalaya Rivers, and the northern Gulf of Mexico.

William Lehr
National Oceanic and Atmospheric Administration

William Lehr received his Ph.D. in Physics from Washington State University in 1976. Dr. Lehr was formerly Spill Response Group Leader and is presently Staff Senior Scientist for the National Oceanic and Atmospheric Administration Hazardous Materials Response Division (NOAA/HAZMAT). Dr. Lehr is an expert in the fate and behavior of oil spills, being past chairman on the International Oil Weathering Committee and author of computer models and many publications on this subject. Dr. Lehr provided technical consultation for a joint NOAA/HAZMAT and U.S. Coast Guard research project into tanker leakage rate for heavy oils, including Orimulsion®.

Malcolm MacKinnon III
MSCL

RAdm. Malcolm MacKinnon III, U.S. Navy, Retired, received his M.S. in Naval Architecture & Marine Engineering from MIT in 1961. He is currently the President of MSCL, Inc. RAdm. MacKinnon is an NAE Member, serves on the Marine Board and was a member on several NRC committees, i.e., Committee for Naval Hydromechanics Science and Technology and Committee on Marine Transportation of Heavy Oil.

James R. Payne
Payne Environmental Consultants, Inc.

James R. Payne received his Ph.D. in Chemistry from the University of Wisconsin - Madison in 1974. Currently, Dr. Payne is the President of Payne Environmental Consultants, Inc., which specializes in oil and chemical pollution studies for government and industry. Over the twenty-five years of his professional career, Dr. Payne has been involved in numerous projects dealing with marine- and-water pollution issues. Dr. Payne was a member of the NRC Committee on Effectiveness of Oil Spill Dispersants.

Gary A. Reiter
Westcliffe Environmental Management

Gary Reiter received his M.A. in Marine Affairs from the University of Rhode Island and a Bachelors Degree from the University of Southern Colorado. He has worked in the pollution response field for the past twenty-five years. In 1991, he retired as a Commander from the U.S. Coast Guard with the majority of his time spent in its Marine Environmental Response Program. He served as both an Executive Officer and a Commanding Officer of the Pacific Strike Team and as an Assistant Branch Chief of the Pollution Response Branch of the U.S. Coast Guard. Since retiring he has worked in industry as a Response Manager for BHP Petroleum, Executive Vice President of O'Brien's Oil Pollution Service, and for the past four years has served as President of his own firm - Westcliffe Environmental Management, Inc.

John N. Sacco
New Jersey Department of Environmental Protection

John Sacco received his M.S. in Ecology from North Carolina State University in 1989. Currently, Mr. Sacco is the natural resource damage assessment coordinator for the Office of National Resource Restoration at the New Jersey Department of Environmental Protection. Mr. Sacco has 13 years of experience in the ecological risk and natural resource damage aspects of environmental regulations. Mr. Sacco has responded to over 30 oil spills and chemical discharges. Part of his duties included coordinating field and laboratory data collection and analysis, negotiating damage settlements with responsible parties, and overseeing restoration of injured resources.

Appendix B

Literature Reviewed by the Committee

AEA Technology. 1999. Landspills of Orimulsion 100 and Orimulsion 400. National Environmental Technology Centre, Oxfordshire, U.K: AEA Technology.

AEA Technology. 1998. Studies on the Spill Behavior of Orimulsion-100 and Orimulsion-400 Using Standard WSL and IFP Tests. AEAT-3481. Oxfordshire, UK: AEA Technology.

AEA Technology. The Use of Pumping and Flotation to Recover Sub-Surface Dispersed Orimulsion-400: Tests at OSRL. Oxfordshire, UK: AEA Technology.

Ahel, M., J. McEvoy, and W. Giger. 1993. Bioaccumulation of the lipophilic metabolites of nonionic surfactants in fresh water organisms. Environmental Pollution 79(3):243-248.

Anderson, J. W. 1986. Predicting the effects of complex mixtures on marine invertebrates by use of a toxicity index. In: Environmental Hazard Assessment of Effluents. Special Publication of the Society of Environmental Toxicology and Chemistry. C. H. Ward and B. T. Walton (eds.), New York: Pergamon Press, pp. 115-122.

Anderson, J. W. 1977. Responses to sublethal levels of petroleum hydrocarbons: Are they sensitive indicators and do they correlate with tissue contamination. In: Fate and Effects of Petroleum Hydrocarbons in Marine Organisms and Ecosystems. D. A. Wolf (ed.), New York: Pergamon Press, pp. 95-114.

Ault, J. S., M. A. Harwell, and V. Myers, (eds.). Comparative Ecological Risk Assessment. Technical Support Document for the Comparison of the Ecological Risks to the Tampa Bay Ecosystem From Spills of Fuel Oil #6 and Orimulsion. Volume I- II. Miami, Florida: University of Miami, Center for Marine and Environmental Analyses, Rosenstiel School of Marine and Atmospheric Science.

Banat, I. M., R. S. Makkar, and S. S. Cameotra. 2000. Potential commercial applications of microbial surfactants. Applied Microbiology and Biotechnology 53(5):495-508.

Barber, L. B. II, J. A. Leenheer, W. E. Pereira, T. L. Noyes, G. A. Brown, C. F. Tabor, and J. H. Writegr. 1995. Organic contamination of the Mississippi River. In: Contaminants in the Mississippi River. R. H. Meade, (ed.), U.S. Geological Survey Circular 1133. Washington, D.C.: U.S. Government Printing Office, pp. 115-135.

Bartnik, F., and K. Kunstler. 1987. Biological effects, toxicology and human safety. In: Surfactants in Consumer Products. New York: Springer Verlag, pp. 475-503.

Benggio, B. 1994. An Evaluation of Options for Removing Submerged Oil Offshore Treasure Island, HMRAD 94-5.
Benke, A. C. 2001. Importance of flood regime to invertebrate habitat in an unregulated river-floodplain ecosystem. Journal of the North American Benthological Society 20(2): 225-240.
Benke, A. C., R. L. Henry, D. M. Gillespie, and R. J. Hunter. 1985. Importance of snag habitat for animal production in Southeastern streams. Fisheries 10(5): 8-13.
Benoit, S. 1994. Orimulsion Trials: July 12-14, 1994, Emergency Services Center, Mulgrave, Nova Scotia, Canada.*
Bishop, W. E., and R. L. Perry. 1981. Development and evaluation of a flow-through growth inhibition test with duckweed (*Lemna minor* L.). In: Toxicology and Hazard Assessment, D. R. Branson and K. L. Dickson, (eds). Aquatic Fourth Conference ASTM STP 737.
Bitor America Corporation. Orimulsion: Clean Power for the Future. Boca Raton, Florida: Bitor America Corporation.*
Bitor America Corporation. 1994. Bitor America Corporation Corporate Response Plan, Volume II, Appendix II. Boca Raton, Florida: Bitor America Corporation.*
Bitor America Corporation. 1994. Summary Report: Orimulsion.*
Bitor America Corporation. 1995. Environmental Review of Orimulsion Production and Transport in Venezuela. Boca Raton, Florida: Bitor America Corporation.*
Bitor America Corporation. 1996. Summary of Studies Conducted on the Environment, Fate, and Effects of Orimulsion. Boca Raton, Florida: Bitor America Corporation.*
Bitor America Corporation. 1997. Orimulsion Containment and Recovery Test Carried Out in Puerto La Cruz, Anzoategui State, Venezuela. Final Report. Version 1.1. Boca Raton, Florida: Bitor America Corporation.*
Bitor America Corporation. 1998. Laboratory Testing on the Dispersibility of 2 Orimulsion Formulations, R.98.14.C. Cedre/Bitor.*
Bitor America Corporation. 1999. Orimulsion Fact Sheet, Bitúmenes Orinoco, S.A (PDVSA-Bitor)*
Bitor America Corporation. 1999. Orimulsion Fact Sheet—Air Pollutant Emissions of Trace Elements and Compounds, Bitúmenes Orinoco, S.A (PDVSA-Bitor)*
Bitor America Corporation. 1999. Orimulsion Fact Sheet—Ecological Risk Assessment, Bitúmenes Orinoco, S.A (PDVSA-Bitor) *
Bitor America Corporation. 1999. Orimulsion Spill Response Manual Volume II: Protection and Cleanup of Marine Shorelines. Boca Raton, Florida: Bitor America Corporation.*
Bitor America Corporation. 1999. Underwater Remote Detection (URD) of Orimulsion Using Sonar, Phase 1 Report. Boca Raton, Florida: Bitor America Corporation.*
Bitor America Corporation.1999. Safety Plan for the Transportation of Orimulsion from New Orleans to Henderson, Kentucky. Orimulsion Barge Safety Plan. Boca Raton, Florida: Bitor America Corporation.*
Bitor Europe Limited. Containment/Recovery/Clean-Up Technology for Orimulsion Spills.*
Bitor Europe Limited. 1994. Orimulsion: A Natural Bitumen-in-Water Emulsion. Health Safety and Environment Manual - Version 4.0.: Bitor Europe Limited.*
Bitor Europe Limited. 1994. Orimulsion Emergency Manual. Boca Raton, Florida: Bitor America Corporation.*
Bitúmenes Orinoco, S.A (PDVSA-Bitor). 1996. Final Report of Orimulsion Containment and Recovery Test Carried Out in Puerto La Cruz, Anzoatequi State, Venezuela. Boca Raton, Florida: Bitor America Corporaton.*
Bitúmenes Orinoco, S.A (PDVSA-Bitor). 1999. Orimulsion Fact Sheet: Ecological Risk Assessment. Boca Raton, Florida: Bitor America Corporation.*
Bitúmenes Orinoco, S.A (PDVSA-Bitor). 1999. Orimulsion Fact Sheet—Fine Particulate Matter Emissions. Boca Raton, Florida: Bitor America Corporation.*
Bitúmenes Orinoco, S.A (PDVSA-Bitor) 1999. Environmental Aspects Associated with the Life Cycle of Orimulsion, Bitúmenes Orinoco, S.A (PDVSA-Bitor).

Bitúmenes Orinoco, S.A (PDVSA-Bitor). 2001. Orimulsion Fact Sheet - Physical and Chemical Properties of Orimulsion-400. Boca Raton, Florida: Bitor America Corporation.*

Boggis, C. J., M. S. Hamilton, and M. J. Herz. 1999. Threading the Needle: Analysis of the *Julie N* Oil Spill and Its Aftermath, Portland, Maine: University of Southern Maine.

Bowadt, S., B. Jensen, and K. Lund. 1998. Development of an Electrospray and Determination of Alcohol Ethoxylates in Water Samples Down to Trace Levels, Hørsholm, Denmark, VKI.

Brey, L., J. Rodríguez Grau, and W. Feragotto. 1995. Toxicity Evaluation of Orimulsion, Bitumen Cerro Negro, Intan 100, and Fuel Oil No. 6 on the Fertility of the Sea Urchin Species *Echinometra lucenter*, and on Survival of the Coral Species *Tubastrea aurea*. Caracas, Venezuela: Intevep, S.A., Center for Research and Technological Support of Petróleos de Venezuela.

Brown, D., H. de Henau, J. T. Garrigan, P. Gerike, M. Holt, E. Keck, E. Kunke, E. Matthjis, J. Waters, and R. J. Watkinson. 1986. Removal of nonionics in a sewage treatment plant. Removal of domestic detergent nonionic surfactants in an activated sludge sewage treatment plant. Tenside 23:190-195.

Bruno, M. S. 1995. Laboratory Study of the Dispersion Characteristics of Orimulsion. Proprietary Technical Report No. SIT-DL-95-9-2734. Hoboken, New Jersey: Stevens Institute of Technology, Davidson Laboratory.

Bryan, C. F., D. A. Rutherford, and B. Walker-Bryan. 1992. Acidification of the lower Mississippi River. Transactions of the American Fisheries Society 121(3):369-377.

Burns, G. H. III, C. A. Benson, T. Eason, J. Michel, S. Kelly, B. Benggio, and M. Ploen. 1995. Recovery of Submerged Oil at San Juan, Puerto Rico 1994. In: Proceedings of the 14th International Oil Spill Conference, Long Beach, CA. pp 551-557

Cardellini, P., and L. Ometto. 2001. Teratogenic and toxic effects of alcohol ethoxylate and alcohol ethoxy sulfate surfactants on *Xenopus laevis* embryos and tadpoles. Ecotoxicology and Environmental Safety 48(2):170-177.

Castle, R. W., F. Wehrenberg, J. Bartlett, and J. Nuckols. 1995. Heavy Oil Spills: Out of Sight, Out of Mind. In: Proceedings of the 14th International Oil Spill Conference, Long Beach, California, pp. 565-571.

Chemical Manufacturers Association (CMA). 1991. Early life stage toxicity of nonylphenol to the fathead minnow, *Pimephales promelas*. Unpublished report. Washington, D.C.: CMA.

Cooper, D., and F. Hvidbak. 2000. Evaluation of mechanical recovery devices for spills of Orimulsion. In: Proceedings of the 23rd Arctic and Marine Oilspill Program, Technical Seminar, 2000. Environment Canada, 1:337-351.

Dolan, J. M. III, C. B. Gregg, J. Cairs Jr., K. L. Dickson, and A. C. Hendricks. 1974. The acute toxicity of three new surfactant mixtures to a mayfly larvae. Archiv fuer Hydrobiologie 74(1):123-132.

Dorn, P. B., J. H. Rodgers, S. T. Dubey, W. B. Gillespie, and R. E. Lizotte. 1997. An assessment of the ecological effects of a C_{9-11} linear alcohol ethoxylate surfactant in stream mesocosm experiments. Ecotoxicology 6(5):275-292.

Dorn, P. B., J. H. Rodgers, Jr., S. T. Dubey, W. B. Gillespie, Jr., and A. R. Figueroa. 1996. Assessing the effects of a C_{14-15} linear alcohol ethoxylate surfactant in stream mesocosms. Ecotoxicology and Environmental Safety 34(2):196-204.

Dunphy, J. C., D. G. Pessler, S. W. Morral, K. A. Evans, D. A. Robaugh, G. Fujimoto, and A. Negahban. 2001. Derivatization LC/MS for the simultaneous determination of fatty alcohol and alcohol ethoxylate surfactants in water and wastewater samples Environmental Science and Technology 35(6):1223-1230.

Energy Information Administration. 1998. International Petroleum Statistics Report, U.S. Government Printing Office, Washington, D.C.

Environment Canada, Emergencies Science Division. 1994. Physical Properties, Chemical Composition, Dispersibility, and Toxicity of Orimulsion. Final Draft. Ottawa, Canada: Environment Canada, Emergencies Science Division.

EPACH Corporation, and KBN Engineering and Applied Sciences. 1995. Acute Toxicity Tests of Orimulsion and Fuel Oil No. 6 Using Standard Test Species, COSAP Technical Document for Licensing Support. Center for Marine and Environmental Analyses, University of Miami, Unpubl.*

Etkin, D. S. 2001. Tanker and Barge Oil Spills in US Waterways 1973-2000. Washington, D.C.: National Research Council, Ocean Studies Board.

Febres, G. A., J. A. Gonçalvez, and J. Vilas. 1995. Fate and Behavior of Orimulsion Spilt in Sea Water, Intevep, S.A.

Federle, T. W., R. M. Ventullo, and D. C. White. 1990. Spatial distribution of microbial biomass, activity, community structure, and the biodegradation of linear alkylbenzene sulfonate (LAS) and linear alcohol ethoxylate (LAE) in the subsurface. Microbial Ecology 20:297-313.

Feijtel, T. C. J., J. Struijs, and E. Matthijs. 1999. Exposure modeling of detergent surfactants—Prediction of 90th-percentile concentrations in the Netherlands. Environmental Toxicology and Chemistry 18(11):2645-2652.

Field Studies Council, and Oil Pollution Research Unit (OPRU). 1991. Orimulsion Spillage Emergency Plan—Mersey Estuary.

Field Studies Council. 1992. Orimulsion Spillage Emergency Plan, Milford Haven, Oil Pollution Research Unit.

Fingas, M. Heavy Oil Behaviour in the Ocean, Ottawa, Canada: Environment Canada.

Frank Ayles and Associates, Ltd. 1995. Report on the Orimulsion Pollution Control Testing Carried Out at OSSC, Southampton, U.K., Ref: FAA/SP102A. London, U.K.: Frank Ayles and Associates, Ltd.

Fremling, C. R., J. L. Rasmussen, S. P. Cobb, C. F. Bryan, R. E. Sparks, R. O. Caflin, and D. O. Editor Dodge. 1989. Mississippi River Fisheries. An International Large River Symposium, Ottawa, Canada: Canadian Special Publication, Fisheries and Aquatic Sciences.

Fujii, T. 1998. Biodetergents. In: New Products and Applications in Surfactant Technology, D. R. Karsa (ed.). Sheffield, U.K.: Sheffield Academic Press, pp. 88-108.

Fujita, M., M. Ike, K. Mori, H. Kaku, Y. Sakaguchi, M. Asano, H. Maki, and T. Nishihara. 2000. Behaviour of nonylphenol ethoxylates in sewage treatment plants in Japan—Biotransformation and ecotoxicity. Water Science and Technology 42(7-8):23-30.

Garcia Tavel, N. 2001. Presentation given at National Research Council Meeting, May 3, 2001, Boca Raton, Florida. In: Orimulsion: Clean Power for the Future, Bitor America Corporation *

Garcia Tavel, N., and I. C. Johnson. 1999. Orimulsion-400, the Next Generation: Environmental Fate, Effects and Recovery. In: Proceedings of the 1999 International Oil Spill Conference: March 8-11, 1999, Seattle, Washington, pp. 1233-1238.

Gillespie, W. B. Jr., J. H. Rodgers Jr., and P. B. Dorn. 1997. Responses of aquatic invertebrates to a C_{9-11} non-ionic surfactant in outdoor stream mesocosms. Aquatic Toxicology 37(2-3):221-236.

Gledhill, W. E., R. L. Huddleston, L. Kravetz, A. M. Nielsen, R. I. Sedlak, and R. D. Vashon. 1989. Treatability of surfactants at a wastewater plant. Tenside 26:276-281.

Golder and Associates, Inc. 1999. Orimulsion-400 and Fuel Oil No. 6: A comparative study of aquatic ecotoxicology (in the Mediterranean Sea), no. 992210/3683. Torino, Italy: Golder Associates.*

Grammo, A., R. Eklund, M. Berggren, and K. Magnusson. 1991. Toxicity of 4-nonyl phenol to aquatic organisms and potential for bioaccumulation. In: Swedish EPA Seminar on Nonylphenol Ethoxylates/Nonylphenol, Saltsjobaden, Sweden, pp. 53-75.

Guenette, C. 1991. Data Report on Behaviour of Orimulsion Spills On Water for Intevep, S.A. Ottawa, Ontario: O.S.L. Ross Environmental Research Limited.

Guénette, C., and G. Sergy. 1999. Disposal Options for Recovered Bitumen, Ottawa, Canada: Environment Canada, Emergencies Science Division.

Guénette, C., and G. Sergy. 1999. Orimulsion Spill Response Manual Volume III: Disposal Options for Recovered Bitumen, Ottawa, Canada: Emergencies Science Division, Environment Canada.

Gunter, P. A. 1990. An assessment of the dispersity of Orimulsion during an Orimulsion spill test in the North Sea. Branch/Project CSB/191. BP Research, Sunbury Research Centre.*

Gunter, P. A. 1991. A laboratory study to investigate the destabilization of Orimulsion observed during spill tests. Branch Report No. 191 075. BP Research, Sunbury Research Centre.*

Hager, C. -D. 1998. Alkylphenol ethoxylates: biodegradability, aquatic toxicity and environmental fate. In: New Products and Applications in Surfactant Technology, D. R. Karsa (ed.). Sheffield, U.K.: Sheffield Academic Press Ltd., pp. 1-29.

Harper, J., G. Sergy, and M. Kory. 1997. Orimulsion Sediment Interaction Scoping Experiments. In: Proceedings of the 20th Arctic and Marine Oilspill Program (AMOP) Technical Seminar - Volume II. Ottawa, Canada, Environment Canada.

Harrelson, R. A., J. H. Rodgers, R. E. Lizotte, and P. B. Dorn. 1997. Responses of fish exposed to a C_{9-11} linear alcohol ethoxylate nonionic surfactant in stream mesocosms. Ecotoxicology 6(6):321-333.

Harwell, M. A., J. S. Ault, and J. H. Gentile. 1995. Comparative Ecological Risk Assessment: Comparison of the Ecological Risks to the Tampa Bay Ecosystem from Spills of Fuel Oil #6 and Orimulsion. In: Orimulsion: Clean Power for the Future, Bitor America Corporation *

Hennepin Power Station. 1997. Facility Response Plan. Hennepin Power Station, Hennepin, Illinois.

Holt, W. 2001. Marine Transportation & Spill Control. Presentation given at National Research Council Meeting on May 3, 2001 as part of overall presentation by Bitúmenes Orinoco, S.A (PDVSA-Bitor) entitled Natural Energy for a Brighter World. Boca Raton, Florida: Bitor America Corporation.*

Huber, M., U. Meyer, and P. Rys. 2000. Biodegradation mechanisms of linear alcohol ethoxylates under anaerobic conditions. Environmental Science and Technology 34:1737-1741.

Hvidbak, F., S. M. Jacobsen, and P. Masciangioli. 2000. Underwater Remote Detection and Monitoring of Spilled Orimulsion Using Sonar. In: Proceedings of the 23rd Arctic and Marine Oil Spill Program Technical Seminar, Ottawa, Canada, Environment Canada, 1:507-18.

HydroQual-Golder Laboratories Ltd. 2000. Toxicity of Orimulsion to the Freshwater Macrophyte, *Lemna minor* (duckweed). Calgary, Alberta, Canada: HydroQual-Golder Laboratories Ltd.

Intevep S.A. 1994. Comparative Oil/Orimulsion Spill Assessment Program (COSAP): Compilation of Environmental Field and Laboratory Studies on Orimulsion, Its Constituents and Fuel Oil No.6, Final Report. Florida Power and Light Company (FPL).

Johnson, I. C. 2001. Environmental Toxicology. Presentation given at National Research Council Meeting on May 3, 2001 in Washington D.C. as part of overall presentation by Bitúmenes Orinoco, S.A (PDVSA-Bitor) entitled Natural Energy for a Brighter World. Boca Raton, Florida: Bitor America Corporation.*

Johnson, I. C. 1998. Summary of the Environmental Fate and Effects of Orimulsion- 400. Gainesville, Florida: Golder Associates, Inc.

Johnson I. C., and D. French McCay. 2001. Freshwater and Estuarine Comparative Impact Assessment. Presentation given at National Research Council Meeting on May 3, 2001 in Washington D.C. as part of overall presentation by Bitúmenes Orinoco, S.A (PDVSA-Bitor) entitled Natural Energy for a Brighter World. Bitor American Corporation: Boca Raton, Florida.*

Johnson, I. C., C. Metcalfe, Y. Kiparissis, G. Balch, S. Ward, J. Wheat, J. Liu, and T. Potter. 1997. Partial Life-Cycle Studies Using the Estuarine Fish Sheepshead Minnow to Evaluate the Potential Reproductive and Estrogenic Effects of Intan-100, Golder Associates, Inc.

Jokuty, P., M. Fingas, B. Fieldhouse, S. Whitcar, and J. Latour. 1998. Bitumen adhesion as a function of time. In: Proceedings of the 21st Arctic and Marine Oilspill Program, Technical Seminar, 1998. Ottawa, Canada: Environment Canada 1:27-33.

Jokuty, P., B. Fieldhouse, M. Fingas, S. Whitcar, and J. Latour. 1998. Characterizing the dynamics of Orimulsion spills in salt, fresh and brackish waters. In: Proceedings of the 21st Arctic and Marine Oil Spill Program Technical Seminar, 1998. Ottawa, Canada: Environment Canada, 1:33-50.

Jokuty, P., S. Whiticar, B. Fieldhouse, Z. Wang, P. Lambert, and M. Fingas. 2000. A Review of Methods Used in the Preparation and Analysis of Orimulsion Related Samples. In: Proceedings of the 23rd Arctic and Marine Oilspill Program (AMOP), Technical Seminar, 261-79Ottawa, Canada: Environment Canada, Emergencies Science Division.

Jonas, K. 1996. Marbach III Power Plant Trial Operation with Orimulsion Report on Analysis and Evaluation of the Emission Measurements.*

Junk, W. J., and G. E. Weber. 1996. Amazonian Floodplains: a Limnological Perspective. Verein. Limnology 26: 149-57.

Kaperick, J. A. 1997. Oil Beneath the Water Surface and Review of Currently Available Literature on Group V Oils: An Annotated Bibliography, Report No. HMRAD 95-8. Seattle, Washington: National Oceanic and Atmospheric Administration.

Karsa, D. R., (ed.) 1998. New Products and Applications in Surfactant Technology. Sheffield, U.K.: Sheffield Academic Press Ltd.

Kent, R.A., D. Andersen, P.Y. Caux, and S. Teed. 1999. Canadian water quality guidelines for glycols—An ecotoxicological review of glycols and associated aircraft anti-icing and deicing fluids. Environmental Toxicology 14(5):481-522.

Khan, Skander. 1996. Orimulsion- Viability as a Repowering Fuel. The 21st International Technical Conference on Coal Utilization and Fuel Systems, 9 Clearwater, Florida.

Kutt, E. C., and D. F. Martin. 1974. Effect of selected surfactants on the growth characteristics of *Gymnodinium breve* Marine Biology 28:253-259.

La Schiazza, J. A., J. Rodríguez Grau, and F. Losada. 1995. Effects and Recovery of Biofouling Communities Impacted by a Controlled Orimulsion Spill, Intevep, Center for Research and Technological Support of Petróleos de Venezuela.

Lapham, L., L. Proctor, and J. Chanton. 1999. Using respiration rates and stable carbon isotopes to monitor the biodegradation of Orimulsion by marine benthic bacteria. Environmental Science and Technology 2035-39

Lee, S. C., D. Mackay, F. Bonville, E. Joner, and W. Y. Shiu. 1989. A Study of the Long-Term Weathering of Submerged and Overwashed Oil, Report no. EE-119. Ottawa, Canada: Environment Canada.

Lee, S. C., W. Y. Shiu, and D. Mackay. 1992. A Study of the Long Term Fate and Behaviour of Heavy Oils, Report no. EE-128. Ottawa, Canada: Environment Canada.

Leitman, P., and L. Proctor. 2000. Biodegradation of Weathered Orimulsion and Fuel Oil #6 by Marine Sediment Bacteria. Tallahassee, Florida: Florida State University.

Lewis, M. A., and B. G. Hamm. 1986. Environmental modification of the photosynthetic response of lake plankton to surfactants and significance to a laboratory-field comparison. Water Research 20:1575-1582.

Lewis, M. A., and D. Suprenant. 1983. Comparative acute toxicities of surfactants to aquatic invertebrates. Ecotoxicology and Environmental Safety 7:313-322.

Idelfonso Liñero A., M. M. Jimenez, V. J. Andrade, M. J. La Schiazza, and J. Rodriguez Grau. 1994. Evaluation of the Impact of Controlled Spills of Orimulsion. In: Sandy Beach Communities within the Area of Jose, Anzoategui State. Technical Report no. INT-02876, 94. Los Teques, Venezuela: Ecological and Environmental Investigations.

Lizotte, R. E. Jr., D. C. L. Wong, P. B. Dorn, and J. H. Rodgers, Jr. 1999. Effects of a homologous series of linear alcohol ethoxylate surfactants on fathead minnow early life stages. Archives of Environmental Contamination and Toxicology 37(4):536-541.

Lorenzo, T. 1996. Orimulsion Containment and Recovery Tests: Trip Report, Puerto La Cruz, Venezuela, Report no. 4808-1. Ottawa, Canada: Environment Canada.

Madai, J., and H. An der Lan. 1964. Zur wirkung einiger detergentien und subwasser-organismen. Wasser Abwasser 4:168-183.

Maki, A. W. 1979. Correlations between *Daphnia magna* and fathead minnow (*Pimephales promelas*) chronic toxicity values for several classes of test substances. Journal of the Fisheries Research Board of Canada 36:411-421.

Maki, A. W., and W. E. Bishop. 1979. Acute toxicity studies of surfactants to *Daphnia magna* and *Daphnia pulex*. Archives of Environmental Contamination and Toxicology 8:599-612.

Maki, A. W., A. J. Rubin, R. M. Sykes, and R. L. Shank. 1979. Reduction of nonionic surfactant toxicity following secondary treatment. Journal of Water Pollution Control Federation 51:2301-2313.

Maki, H., N. Masuda, Y. Fujiwara, M. Ike, and M. Fujita. 1994. Degradation of alkyphenol ethoxylates by *Pseudomonas* sp. strain TR01. Applied and Environmental Microbiology 60(7):2265-2271.

Maki, H., H. Okamura, I. Aoyama, and M. Fujita. 1998. Halogenation and toxicity of the biodegradation products of a nonionic surfactant, nonylphenol ethoxylate. Environmental Toxicology and Chemistry 17:650-654.

Mann, R.M., and J. R. Bidwell. 2001. The acute toxicity of agricultural surfactants to the tadpoles of four Australian and two exotic frogs. Environmental Pollution 114:195-205.

Mann. A.H., and V.W. Reid. 1971. Biodegradation of synthetic detergents evaluation by community trials. Part 2. Alcohol and alkylphenol ethoxylates. Journal of the American Oil Chemists Society 48:794-797.

Marine Spill Response Corporation. 1993. MSRC Workshop Report: Research on the Ecological Effects of Dispersants and Dispersed Oil, 93-104. Washington, D.C.: Marine Spill Response Corporation.

Masciangioli, P., M. Romero, and M. Farias. 1995. Orimulsion Spill Tests in Low Temperature Environments, Report no. INT-EAG-0001, 95. Caracas, Venezuela: Intevep, S.A.

Matthijs, E., M. S. Holt, A. Kiewiet, and G. B. J. Rijs. 1999. Environmental monitoring for linear alkylbenzene sulfonate, alcohol ethoxylate, alcohol ethoxy sulfate, alcohol sulfate, and soap. Environmental Toxicology and Chemistry 18(11):2634-2644.

Mayer, L., Z. Chen, R. Findlay, J. Fang, S. Sampson, L. Self, P. Jumars, C. Quetél, and O. Donard. 1996. Bioavailability of sedimentary contaminants subject to deposit-feeder digestion. Environmental Science and Technology 30:2641-2645.

Mayer, L. M., L. L. Schick, R. F. L. Self, P. A. Jumars, R. H. Findlay, Z. Chen, and S. Sampson. 1997. Digestive environments of benthic macroinvertebrate guts: Enzymes, surfactants, and dissolved organic matter. Journal of Marine Research 55:785-812.

McAvoy, D. C., S. D. Dyer, N. J. Fendinger, W. S. Eckhoff, D. L. Lawrence, and W. M. Begley. 1998. Removal of alcohol ethoxylates, alkyl ethoxylate sulfates, and linear alkybenzene sulfonates in wastewater treatment. Environmental Toxicology and Chemistry 17(9):1705-1711.

McDonald, J. L. 1995. The *Morris J. Berman* Spill - MSRC's offshore operations. In: Proceedings of the 14th International Oil Spill Conference, Long Beach, California. Washington, D.C.: American Petroleum Institute, pp. 701-706.

Michel, J., and J. A. Galt. 1995. Conditions Under Which Floating Slicks Can Sink in Marine Settings. In: Proceedings of the 14th International Oil Spill Conference, Long Beach, California. Washington, D.C.: American Petroleum Institute, pp 574-576.

Michel, J., D. Scholz, C. B. Henry, and B. L. Benggio. 1994. Group V Fuel Oils: Source, Behavior, and Response Issues. Seattle, Washington: Hazardous Materials Response and Assessment Division, National Oceanic and Atmospheric Administration.

Mitsubishi Corporation. 1990. Final Report on MHI's Study on the Behavior of Orimulsion in Sea Water. Bitor, S. A.*

Mitsubishi Corporation. 1993. Orimulsion Underwater Behavior Test: Test Report. Ship & Ocean Financial Group, Tsukuba Laboratory.*

Møller, L., C. Helweg, F. Pedersen, and E. Bjørnestad. 1999. Environmental hazard profiles of Orimulsion 400 and Heavy Fuel Oil, VKI Project No. 11896. Hørsholm, Denmark: VKI, Institute for the Water Environment.
Moore, S. B., R. A. Diehl, J. M. Barnhardt, and G. B. Avery. 1987. Aquatic toxicities of textile surfactants. Textile Chemist and Colorist 19:29-32.
Morgan, D. C., and W. A. Fernie. 1995. Report on the Orimulsion Pollution Control Testing Carried Out. OSSC, Southhampton, U.K.: Frank Ayles & Associates Limited.
Muller, M. T., A. J. B. Zehnder, and B. I. Escher. 1999. Membrane toxicity of linear alcohol ethoxylates. Environmental Toxicology and Chemistry 18(12): 2767-2774.
National Orimulsion Spill Plan for Lithuania Manual.*
Neff, J. M., B. A. Cox, D. Dixit, and J. W. Anderson. 1976. Accumulation and release of petroleum-derived aromatic hydrocarbons by four species of marine animals. Marine Biology 38:279-289.
New Brunswick Power Dalhousie Generating Station. 1995. Orimulsion Spill Contingency Plan, New Brunswick, Canada.
Oil Spill Response Limited, AEA Technology, Steptech Instruments, and Bitor Europe Limited. 1995. Orimulsion Spill Trials; Recovery and Cleanup of Bitumen, Detention of Sub Surface Plume. Bitor Europe, Ltd.*
Oil Spill Response Limited. 1989. A Report on a Study to Determine Treatment Options Following Spillage of Orimulsion into Marine and Freshwater Environments, South Hampton, U.K.: Oil Spill Service Center.
Ostazeski, S. A., S. P. Daling, S. C. Macomber, D. W. Fredriksson, G. S. Durrell, A. D. Uhler, M. Jones, and K. Bitting. Weathering Properties and The Predicted Behavoir at Sea of Lapio Oil (Weathered No. 6 Fuel Oil). Duxbury, MA: Battelle, Ocean Sciences.
Pang, S. N. J. 1993. Final report on the safety assessment of Polyethylene Glycols (PEGs) -6, -8, -32, -75, -150, -14M, -20M. Journal of the American College of Toxicology 12(5):429-457.
Patoczka, J., and G. W. Pulliam. 1990. Biodegradation and secondary effluent toxicity of ethoxylated surfactants. Water Research 24:965-972.
Payne, A. G., and R. H. Hall. 1979. A method for measuring algal toxicity and its applications to the safety assessment of new chemicals. In: Aquatic Toxicology, L. L. Marking and R. A. Kimerle, (eds). ASTM STP 667. Pp. 171-180.
Petroleos de Venezuela, S.A., (Bitor). 2000. Guidelines for Orimulsion Marine Terminals Spill Response Equipment Package.*
Petroleos de Venezuela, S.A. (Bitor). 1999. Orimulsion Spill Response Manual, Volume III, Disposal Options for Recovered Bitumen.*
Petroleos de Venezuela, S.A. (Bitor). 1999. Orimulsion Spill Response Manual. Boca Raton, Florida: Bitor America Corporation.*
Pedersen, F., J. Larsen Torp, P. Dørge, M. B. Andersen, M. J. Lintrup, L. Møller, and D. Rasmussen. 1999. Environmental Spill Modeling and Risk Assessment of Orimulsion-400 and Heavy Fuel Oil, 11896. Hørsholm, Denmark: VKI, Institute for the Water Environment.
Potter, T., E. Calabrese, P. Kostecki, K. Simmons, and J. Wu. 1997. Chemical Characteristics of the Water Soluble Fraction of Orimulsion-in-Water Dispersions. Amherst, Massachusetts: University of Massachusetts, Department of Food Science and School of Public Health,
Potter, T. L., and B. Duval. 2001. Cerro Negro bitumen degradation by a consortium of marine benthic microorganisms. Environmental Science & Technology 35(1):76-83.
Quilici, A., P. Vásquez, C. Infante, J. Rodríguez-Grau, J. La Schiazza, H. Briceno, and N. Pareira. 1995. Comparative Effects of Spilt Orimulsion, Fuel Oil No.6 and Lago Medio Crude Oil Over Some Functional Variables of an Estuarine Mangrove Ecosystem. Caracas, Venezuela: Intevep, S.A.
Rand, G. M., (ed.). 1995. Fundamentals of Aquatic Toxicology Effects, Environmental Fate, and Risk Assessment, 2nd ed. Bristol, Pennsylvania: Taylor & Francis.

Reilly, T. J., N. C. Kraus, W. R. Wise, and R. Jamail. 1994. Coastal Oilspill Simulation System Prototype Testing Program, 94-005. Washington, D.C.: Marine Spill Response Corporation.

Rosen, M. J., F. Li, S. W. Morrall, and D. J. Versteeg. 2001. The relationship between the interfacial properties of surfactants and their toxicity to aquatic organisms. Environmental Science and Technology 35: 954-959.

Routledge, E. J., and J. P. Sumpter. 1996. Estrogenic activity of surfactants and some of their degradation products assessed using a recombinant yeast screen. Environmental Toxicology and Chemistry 15:241-248.

Rowland, S., P. Donkin, E. Smith, and E. Wraige. 2001. Aromatic hydrocarbon "humps" in the marine environment: Unrecognized toxins? Environmental Science and Technology 35: 2640-2644.

S.L. Ross Environmental Research Ltd. 1998. Identification of Oils that Produce Non-buoyant In Situ Burning Residues and Methods for Their Recovery, Ottawa, Canada.

Salanitro, J. P., and L. A. Diaz. 1995. Anaerobic biodegradability testing of surfactants. Chemosphere 30(5):813-830.

Salanitro, J. P., G. C. Langston, P. B. Dorn, and L. Kravetz. 1988. Activated-sludge treatment of ethoxylate surfactants at high industrial use concentrations. Water Science and Technology 20(11-12):125-130.

Salloum, M. J., M. J. Dudas, W. B. McGill, and S. M. Murphy. 2000. Surfactant sorption to soil and geologic samples with varying mineralogical and chemical properties. Environmental Toxicology and Chemistry 19(10):2436-2442.

Scholz, D. K., J. Michel, C. B. Henry, and B. Benggio. 1994. Assessment of Risks Associated With the Shipment and Transfer of Group V Fuel Oil, 94-8. Seattle, Washington: National Oceanic and Atmospheric Administration.

Scorpio Ship Management. 1996. Scorpio Ship Management Shipboard Oil/Orimulsion Pollution. Boca Raton, Florida: Bitor America Corporation.*

Scott, G. I., T. G. Ballou, and J. A. Dahlin. 1984. Summary and Evaluation of the Toxicological and Physiological Effects of Pollutants on Shellfish-Part 2: Petroleum Hydrocarbons: Report No. 84-31. Columbia, South Carolina: Research Planning, Inc.

Scott, D. 1993. Technical Evaluation of the Coastal Oilspill Simulation System Prototype. 93-029. Washington, D.C.: Marine Spill Response Corporation.

Scott, W. G., and J. Lui Hui. Acute Toxicity of Orimulsion and Orimulsion 400 to Saltwater Organisms. Jupiter, Florida: Toxicon Environmental Sciences.

Seakem Oceanography Ltd. 1986. Oceanographic Conditions Suitable for the Sinking of Oil, EPS 3/SP/2. Sidney, British Columbia: Seakem Oceanography Ltd.

Sharer, D. H., L. Kravetz, and J. B. Carr. 1979. Biodegradation of nonionic surfactants. Tappi 62(10):75-78.

Shell Chemical Company. 1988. (Unpubl.) Aquatic Toxicity of Neodol* 1-5 to Algae, Daphniids and Fish. Technical Progress Report WRC 225-87. Houston, Texas: Shell Development Company.

Shell Oil Company. 1983. Chemical Economics Handbook. Menlo Park, Califorinia: Shell Oil Company.

Simececk-Beatty, D. A., W. J. Lehr, and R. A. Jones. 1998. Modeling the Release, Behavior, and Fate of Bitumen—Water Emulsions. In: Proceedings of the 21st Arctic and Marine Oilspill Program (AMOP), Technical Seminar, Ottawa, Canada: Environment Canada, Environmental Services Division. Pp. 439-447

SK Energy Asnaes Power Plant. 1999. Spills Contingency Plans for Orimulsion at Asnaevaerket, Bitor Europe.*

S.L. Ross Environmental Research Ltd., Ottawa, Ontario. 1987. The Transient Submergence of Oil Spills: Tank Tests and Modeling, Environment Canada EE-96.*

Sommerville, M. 1996. Investigations into Freshwater Spills of Orimulsion, AEA/WMES/RMAF/@0206001/R/02. Oxfordshire, U.K.: AEA Technology.

Sommerville, M. 1990. (Unpubl.) Observations of the Behavior of Orimulsion Released into the Sea. CR 3361 (MPBM). Hertfordshire, U.K.: Warren Spring Laboratory.

Sommerville, M., T. Lunel, N. Bailey, D. Oland, C. Miles, P. A. Gunter, and T. Waldhoff. 1987. Orimulsion. In: Proceedings of the 1987 Oil Spill Conference (prevention, behavior, control, cleanup) April 6-9, 1987, Baltimore, Maryland. Washington, D.C.: American Petroleum Institute, pp. 7479-7483.

Soto, A. M., H. Justicia, J. W. Wray, and C. Sonnenschein. 1991. Para-nonyl-phenol: An estrogenic xenobiotic released from modified polystyrene. Environmental Health Perspectives 92:167-173.

Stanton, E. M. Operational Considerations—Tank Barge Morris J. Berman Spill. Old San Juan, Puerto Rico: Marine Safety Office.

Steber, J., and P. Wierich. 1987. The anaerobic degradation of detergent range fatty alcohol ethoxylates. Studies with ^{14}C-labeled model surfactants. Water Research 21(6):661-667.

Stout, S. A., P. J. Barrett, and L. G. Roberts. Predicting the Spill Behavior of Orimulsion Using Design of Experiment Principles. In: Proceedings of the 22nd Arctic and Marine Oilspill Program (AMOP), Technical Seminar. Ottawa, Canada: Environment Canada.

Suter, G. W. II. 1995. Introduction to ecological risk assessment for aquatic toxic effects. In Fundamentals of Aquatic Toxicology Effects, Environmental Fate, and Risk Assessment, 2nd Ed., Rand, G. M., (ed.) Bristol, Pennsylvania: Taylor & Francis, pp. 803-816

Roseth, S., T. Edvardsson, T. M. Botten, J. Fuglestad, F. Fonnum, and J. Stenersen. 1996. Comparison of acute toxicity of process chemicals used in the oil refinery industry, tested with the diatom *Chaetoceros gracilis*, the flagellate *Isochrysis galbana*, and the zebra fish, *Brachydanio rerio*. Environmental Toxicology and Chemistry 15(7):1211-1217.

Swedmark, M., B. Braaten, E. Emanuelsson, and A. Granmo. 1971. Biological effects of surface active agents on marine animals. Marine Biology 9:183-210.

Swim, D. Esq., and K. Kennedy, Esq. Florida Power and Light Company vs. State of Florida, Siting Board; Manatee County. District Court of Appeal First District, State of Florida.

Sykes, R. M., A. J. Rubin, S. A. Rath, and M. C. Chang. 1979. Treatability of a nonionic surfactant by activated-sludge. Journal of the Water Pollution Control Federation 51:71-77.

T.A. Herbert and Associates. 1990. An Analysis of the Environmental Risks from Potential Spills Associated with the Shipment of Orimulsion to Sanford, Florida on the St. Johns River, Final Report. Florida Power & Light Company.

Tanaka, T., K. Yamada, T. Tonosaki, T. Konishi, H. Goto, and M. Taniguchi. 2000. Enzymatic degradation of alkylphenols, bisphenol A, synthetic estrogen and phthalic ester. Water Science and Technology 42(7-8):89-95.

Turner, A. H., F. S. Abram, V. M. Brown, and H. A. Painter. 1985. The biodegradability of two primary alcohol ethoxylate nonionic surfactants under practical conditions, and the toxicity of the biodegradation products to rainbow trout. Water Research 19:45-51.

Thomas, A. M. 1994. MSRC Workshop Report: Research on Waterbird Deterrents for Marine Oil Spills, 94-006. Denver, Colorado: Marine Spill Response Corporation.

Tolls, J., M. Haller, E. Labee, M.Verweij, and D. T. H. M. Sijm. 2000. Experimental determination of bioconcentration of the nonionic surfactant alcohol ethoxylate. Environmental Toxicology and Chemistry 19(3):646-653.

United States Coast Guard. 1996. Coast Guard Authorization Act of 1996—Conference Report.

United States Coast Guard, Seventh District. 1995. Group V Petroleum Oils, USCG Seventh District Work Group Report.

van Emden, H. M., C. C. M. Droon, E. N. Schoeman, and H. A. Van Seventer. 1974. The toxicity of some detergents tested on *Aedes aegypti* L., *Lebistes reticulates* Peters, and *Biomphalaria glabrata* Say. Environmental Pollution 6:297-308.

van Ginkel, C. G. 1996. Complete degradation of xenobiotic surfactants by consortia of aerobic microorganisms. Biodegradation 7(2):151-164.

Van Vleet, E. 1997. Review of Literature Related to Research Conducted on Orimulsion: Final Report, Florida Power and Light Company (FPL).
Vashon, R. D., and B. S. Schwab. 1982. Mineralization of linear alcohol ethoxylates and linear alcohol ethoxy sulfates at trace concentrations in estuarine water. Environmental Science and Technology 16:433-436.
VKI. 1999. Orimulsion 400 and Heavy Fuel Oil, Study #11896. Hørsholm, Denmark: VKI.
VKI. 1998. Acute Toxicity Test of Orimulsion 400 with the Crustacean *Acartia Tonsa* (Test Period February 2, 1998–February 20, 1998), GLP Study # 81084-02/067. Hørsholm, Denmark:VKI.
VKI. 1998. Acute Toxicity Test of Orimulsion 400 with Crustacean *Acartia Tonsa* (Test Period March 16, 1998–March 18, 1998), GLP Study # 81044-01/067. Hørsholm, Denmark: VKI.
VKI. 1998. Acute Toxicity Test of Orimulsion 400 with Crustacean *Acartia Tonsa*. Test Period March 17, 1998 - May 8, 1998, 81078/067. Hørsholm, Denmark: VKI, Department of Ecotoxicology.
VKI. 1998. Chronic Toxicity Test of Orimulsion 400 with the Crustacean *Acartia Tonsa*, GLP Study No. 81078/067. Hørsholm, Denmark: VKI.
VKI. 1998. Development of an Electrospray LC-MS Method for the Characterization and Determination of Alcoholethoxylates in Water Samples Down to Trace Levels, 11020. Hørsholm, Denmark: VKI, Institute of the Water Environment.
VKI. 1998. Ecotoxicological characterization of Orimulsion 400, 11020. Hørsholm, Denmark: VKI.
VKI. 1997. Biodegradability of 75090 Monoethalomine - Marine Closed Bottle Test (Test Period October 10, 1997 - November 11, 1997), GLP Study 81047-2/064. Hørsholm, Denmark: VKI.
VKI. 1997. Biodegradability of GENAPOL X159-Marine Closed Bottle Test (Test Period September 30, 1997 - October 28, 1997), GLP Study 81047-02/065. Hørsholm, Denmark: VKI.
Wagner, R. 1978. Behavior of MBAS and BIAS in a municipal sewage treatment plant. Gas-Wasserfach Wasser Abwasser 119:235-242.
Walker, A. H., J. Michel, G. Canevari, J. Kucklick, D. Scholz, C. A. Benson, E. Overton, and B. Shane. Chemical Oil Spill Treating Agents: Herding Agents, Emulsion Treating Agents, Solidifiers, Elasticity Modifiers, Shoreline Cleaning Agents, Shoreline Pre-treatment Agents, and Oxidation Agents, 93-015. Washington, D.C.
Wilson, D., Y. C. Poon, and D. Mackay. 1986. An Exploratory Study of the Buoyancy Behavior of Weathered Oils in Water, EE-85. Toronto, Ontario: Environment Canada.
Wolf, D. A., (ed.) 1977. Fate and Effects of Petroleum Hydrocarbons in Marine Organisms and Ecosystems. New York: Pergamon Press.
Wong, D. C. L., P. B. Dorn, and E. Y. Chai. 1997. Acute toxicity and structure-activity relationships of nine alcohol ethoxylate surfactants to fathead minnow and *Daphnia magna*. Environmental Toxicology and Chemistry 16(9):1970-1976.
Wood, P. 1996. Investigations into Landspills of Orimulsion, AEA/20206001. Oxfordshire, U.K.: AEA Technology.
Yamane, A., M. Okada, and R. Sudo. 1984. The growth-inhibition of planktonic algae due to surfactants used in washing agents. Water Research 18:1101-1105.
Yeh, D. H., K. D. Pennell, and S. G. Pavlostathis. 1998. Toxicity and biodegradability screening of nonionic surfactants using sediment-derived methanogenic consortia. Water Science and Technology 38(7):55-62.

Zeeman, M. G. 1995. Ecotoxicity testing and estimation methods developed under Section 5 of the Toxic Substances Control Act (TSCA). In: Fundamentals of Aquatic Toxicology Effects, Environmental Fate, and Risk Assessment, 2nd Ed., Rand, G. M, (ed.) Bristol, Pennsylvania: Taylor & Francis. Pp. 703-715.

---

[1] Orimulsion® is a registered trademark belonging to Bitúmenes Orinoco, S.A (PDVSA-Bitor) and licensed to Bitor America Corporation.
* Contact Bitor America Corporation at the address below to obtain technical reports.

Bitor America Corporation
5100 Town Center Circle, Suite 301
Boca Raton, Florida 33486
(561) 392-0026
(561) 392-0490
http://www.bitoramerica.com/

Appendix C

Acronyms

AE	alcohol ethoxylate
AEA	Short for a company called AEA Technology
AMOP	Arctic and Marine Oilspill Program
APE	alkylphenol ethoxylates
API	American Petroleum Institute measurement unit
ASTM	American Society of Testing and Materials
BAC	Bitor America Corporation
Bitor	Bitúmenes Orinoco, SA
BTEX	benzene, toulene, ethyl benzene, and xylenes
BTU	british thermal unit
CFT	Clean Fuels Technology
COSAP	Comparative Oil/Orimulsion Spill Assessment Program
DESMI	Name of a pump skimmer manufacturer
DOE	Department of Energy
EIA	Energy Information Administration
EO	ethylene oxide
EPA	Environmental Protection Agency
FAF	Forced Adhesion and Floatation

HMRAD	Hazardous Materials Response and Assessment Division
LAPIO	Low API oil, where API (above) is a measuring unit
LAS	linear alkyl benzene sulfonate
LC-MS	Liquid Chromatograph-Mass Spectrometer
LOEC	Lowest-Observed-Effects Concentration
MAH	Monocyclic Aromatic Hydrocarbon(s)
MEA	monoethanolamine
MIT	Massachusetts Institute of Technology
MSRC	Marine Spill Response Corporation
NAE	National Academy of Engineering
NFPA	National Fire Protection Association
NOAA	National Oceanic and Atmospheric Administration
NOAA/HZMAT	National Oceanic and Atmospheric Administration Hazardous Materials Response Division
NOEC	No Observable Effect Concentration
NPE	alkyl phenyl ethoxylate
NPE	nonylphenol ethoxylate
NRC	National Research Council
OPA-90	Oil Pollution Act of 1990
OPA	Oil Pollution Act of 1990
OPRU	Oil Pollution Research Unit
OSRL	Oil Spill Response Limited (has since changed its name to OSSC)
OSSC	Oil Spill Service Center
PAHS	polynuclear aromatic hydrocarbons
PDVSA	Petroleos de Venezuela, SA
PEG	polyethylene glycols
SMART	Surveillance and Monitoring for Alternative Response Technology
SPM	suspended particular material
TSCA	Toxic Substances Control Act
URD	Underwater Remote Detection
USCG	United States Coast Guard

VKI   Name of consulting firm in Hørsholm, Denmark

WAF   water accommodated fraction
WSF   water-soluble fractions
WSL   Warren Springs Laboratory (since changed name to AEA Technology)